RECENT PROGRESS *in* ENVIRONMENTAL BIOLOGY

Recent Progress *in* Environmental Biology

Dr. B.M. Sharma

Oxford Book Company

Jaipur ▪ New Delhi

Oxford Book Company

Regd. Off. "Bony Residency", Gate No. 2
Opp. Tilak Public School, Vishweswariya Nagar,
Gopalpura Road, Jaipur - 302018, India
Ph: 0141 - 2761280, Telefax: 2761381
email - oxfordbookjpr@gmail.com
website - oxfordbookcompany.com

Delhi Office: 102, 1st Floor, Satyam House,
4327/3, Ansari Road, Darya Ganj,
New Delhi - 110002, India
Ph: 011-45652440

Edition, 2025

ISBN : 9789350300053

Price : ₹ 2500.00

Typeset by Shivangi Computers, Jaipur

Printed and bound in India by Thomson Press Ltd.,
New Delhi

Foreword

There already exist several excellent text books of Environmental Science which are authored by many reputed Indian and foreign writers. But this is an age of continuing education, the efforts to update this treasure of knowledge would continue and would offer more imprints of higher knowledge in more lucid and simple language for the benefit of students and teacher community.

I have pleasure to foreword this book in which Dr. B.M. Sharma has made concerted efforts to compile the environmental issues in the form of a book entitled "Recent Progress in Environmental Biology". It is undoubtedly to say that authors have provided latest and comprehensive knowledge on various aspects of Environmental Biology.

The present book is a sincere and novel attempt in the field of Environmental Science. I am sure that this book will be of immense value of readers.

Prof. Ramesh Chandra

Preface

The present book is intended for the students of Undergraduate, Postgraduate and higher studies in the universities and colleges, as well as those are interested in Environmental Education. This book has current information on various essential aspects of Environmental Biology. An endeavour has been made to illustrate each aspect with comprehensive text, suitable diagrams and update knowledge to guide the students and extension workers in lucid language.

The editor is greatly indebted to all the authors and offers his heartfelt thanks for their contribution in the book. I commend the fruits of my labour to the fraternity of Environmental Biology.

Dr. B.M. Sharma

List of Contributors

Anoop Singh
Environmental Research Institute, University College Cork, Cork, Ireland.

Asheesh Kumar
Mewar University, Chittorgarh, India.

B.M. Sharma
Mewar University, Chittorgarh (Rajasthan).

B.V. Kabra
Department of Chemistry, M.L.V.Govt. College, Bhilwara (Raj.).

Chena Ram
Animal Behaviour Unit, Department of Zoology, J.N.V. University Jodhpur- 342005 (Rajasthan).

Deepa B.
Wood Bio-degradation Division. Institute of Wood Science & Technology, Malleswaram, Bangalore.

Deepa Katiyar
Sikkim Manipal University Extension Centre South of Saket, New Delhi, India.

Dmitro M. Grodzinsky
Professor, Dr Sci, Full Member of the National Academy of Sciences of Ukraine Institute of Cell Biology and Genetic Engineering, Kiev, Ukraine.

G. Chandra
Mosquito and Microbiology Research Units, Department of Zoology, The University of Burdwan, Burdwan – 713104.

Goutam Sharma
Animal Behaviour Unit, Department of Zoology, J.N.V. University Jodhpur- 342005 (Rajasthan).

I. Bhattacharjee
Mosquito and Microbiology Research Units, Department of Zoology, The University of Burdwan, Burdwan – 713104.

L.S. Rajpurohit
Animal Behaviour Unit, Department of Zoology, J.N.V. University Jodhpur- 342005 (Rajasthan).

M.S. Paul
Department of Botany, St. John's College, Agra-282 002, India.

Mayank Varun
Department of Botany, St. John's College, Agra-282 002, India.

Menka Surana
Department of Applied Chemistry, Mewar University, Chittorgarh (Raj.).

N. Kumara Swamy
JSS Acdemy of Technical Education, Noida (UP).

Pratibha Singh
JSS Acdemy of Technical Education, Noida (UP).

Remadevi O.K.
Scientist F and Head, Wood Bio-degradation Division. Institute of Wood Science & Technology, Malleswaram, Bangalore.

Rohan D'Souza
Department of Botany, St. John's College, Agra-282 002, India.

S. K. Starling
Mewar University, Chittorgarh, India.

S.B. Agrawal
Botany Department, Banaras Hindu University, Varanasi, India.

S.K. Chatterjee
Mosquito and Microbiology Research Units, Department of Zoology, The University of Burdwan, Burdwan – 713104.

Contents

1

Reclamation and Management of Saline and Alkali Soils

Anoop Singh, Deepa Katiyar and
S.B. Agrawal

Abstract

Soil is the most precious natural resource and thus requires proper management. Estimates show that the world as a whole is losing at least 3.0 ha of arable land every minute due to salinization or sodification. In India about 7.0 M ha land is affected by salinity and alkalinity. The problem of saline and alkali soils is old but its magnitude and intensity have been increasing because of poor land and water management practices. The proper land management by way of its reclamation involves physical, chemical and biological means, which are site specific and their integration is highly desirable to combat with the problem. The present review is an attempt to emphasize the problem of salinity and alkalinity of soils, its effect on plants and application of physical, chemical and biological methods of soil reclamation along with management issues.

Key Words: Saline soil, alkali soil, reclamation, management, constraints.

Introduction

The formation of one-inch soil takes about 400 years but it degrades in few years only. As a result of our ever-increasing population, the demand for food is increasing day by day, for this we have two options either to increase the land area or to produce more food per unit area. The later option can be exercised but only up to a certain extent. As for the first option the land area is already finite. National waste lands development board estimated that about 175 M ha out of 329 M ha of total land of India belonged to the category of waste lands which yielded zero to a very nominal level, out of which 7Mha is saline or alkali, and the largest area of about 1.25 M ha is found in U.P., which comes about 4 percent of its total geographical land. So we can increase the production by reclaiming such problematic soils. This can be done by various land reclamation practices some are traditional and some modern and it is equally important to judiciously manage the soil after reclamation.

Soils that contain excess soluble salts and affect plant growth adversely are called salt affected soils. These soils are mainly divided in to two major categories.

1. Saline soil: Saline soils are those which contain sufficient soluble salts (chlorides and sulphates of sodium) to interfere with the growth of plants. These soils are locally called as *Reh*.The six classes of saline soils found in India are (1) saline soils (illitic) of semiarid regions with highly saline ground water (2) saline soils (illitic)of sub humid region with moderately saline groundwater (3) saline soils (illitic) of sub humid region (4) saline soils (montmorillonitic) of semi arid delta region (5) saline acid sulfate soils of humid region and (6) saline marsh of arid region.

2. Alkali soils: These are also called as sodic soils and these soils contain excessive exchangeable sodium to adversely affect crop production and locally called as *User*. The alkali soils found in India fall under three classes, viz. (1) alkali soils (illitic) of Indo Gangatic plain with sweet ground water, (2) alkali soils (illitic) of Indo Gangatic plain with sodic ground water and (3) alkali soils (montmorillonitic) of deccan plateaus with sodic or saline ground water.

Balba (1995) compared the saline and alkali soils with normal soil and conclude that pH and EC of saline soil is more than the normal soil due to presence of more sodium salts and alkali soils have more pH and SAR than the normal soils due to presence of excessive exchangeable sodium salts in the soil (Table 1). Properties of these problematic soils are summarized in Table 2.

Table 1

Properties of normal soils compared with saline and alkali soils (Source: Brady, 1996).

Soil	pH	EC (dSM^{-1})	SAR
Normal	6.5	< 4	<15
Saline	7.3-8.5	>4	<15
Alkali	> 8.5	<4	>15

Table 2

Properties of saline and alkali soils. (Source: Abrol, 1998)

Properties	Saline soil	Alkali soil
1. Chemestry of soil solution	Dominated by chlorides and sulphides of sodium, calcium and magnesium.	Dominated by sodium carbonate or sodium bicarbonate or both.
2. Main adverse effect on plants	High osmotic pressure of soil solution.	Alkalinity of soil solution and its corrosive effect on plant roots.
3. Aim of reclamation	Removal of excessive electrolytes through leaching.	Lowering or neutralizing high pH through chemical amelioration.

Factors Responsible for Formation of Saline and Alkali Soils:

1. **Arid and semiarid climate** – In these climatic regions annual evaporation is more than the annual rainfall, resulting which salts present in the lower strata comes on the surface with evaporating water and accumulated on the surface.
2. **Poor drainage of soil-** Due to evaporation of stagnated water in poor drainage areas salts are accumulated on the surface of soil.
3. **High water table** – In high water table areas submerged conditions prevails after evaporation of water such soil becomes saline.
4. **Over flow of seawater over lands** – Seawater is brackish in nature and when it over flow on land, which ultimately increase the concentration of salts.
5. **Bad quality of irrigation water** – Soils become saline when farmers use bad quality water in irrigation, which contains salts in higher concentrations.
6. **Salts blown by wind** – In highly arid region those like Rajasthan, salts are present on the soil surface and blown by high velocity wind, which are deposited on another place.
7. **Saline nature of parent rock material** – When parent rock is saline, soil formed from these rocks is naturally saline.
8. **Continuous use of basic fertilizer** – Continuous application of basic fertilizer increases salts in the field, and makes them saline.
9. **Low permeability of soil to water** – Continuous ploughing at a certain depth forms a hard layer, which restricts the downward movement of water due to which salts are accumulated on surface by evaporation.
10. **Undulated land** – In undulated region water containing salts flows down from upper areas to lower areas, which increases more salts in lower areas.
11. **Presence of hard kanker pan in lower surface** – Due to presence of hard kanker pan in lower surface down word movement of water is inhibited resulting which salts are accumulated on surface by evaporation.

Sources of Salts in Soil

According to Balba (1995) sources of salts are mainly divided in to primary and secondry, primary sources includes parent rocks, marine source and deltaic source, while secondary sources include deforestration, overgrazing, disposal of brakish water, irrigation with salty water, inadequate drainage, etc. Srivastava and Srivastava (1993) reported that organic carbon decreases with increasing soil pH. Organic carbon and nitrogen have a positive correlation that's why as soil pH decreases nitrogen is also decreases (Fig. 1), resulting which soil become less fertile or unfertile.

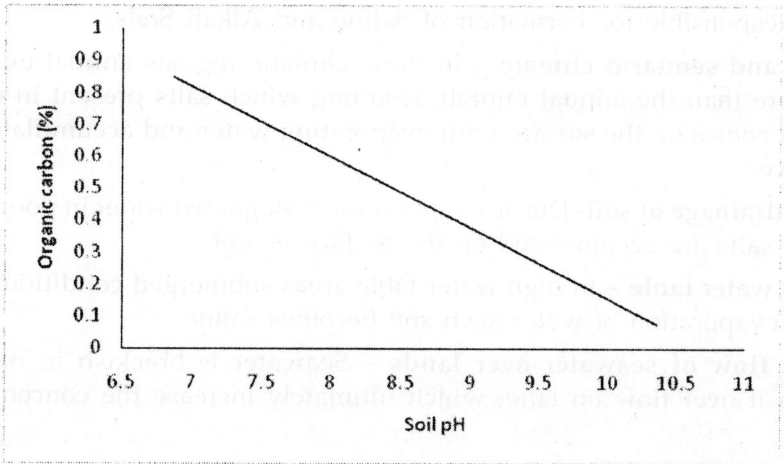

Figure 1: Effect of pH on soil organic carbon (Srivastava and Srivastava, 1993).

Effects of Salts on Soil

When salts are accumulated in the soil (surface and/ subsurface), they reduces the availability of nutrients to plants, as they binded with them. Salts also decreases the activity of microorganisms. Salts are accumulated in subsurface soil and form hard pan, which restrict water to percolate at lower strata. As salts enriches in the soil, the physical properties (texture, porocity, bulk density, etc.) and chemical properties (pH, EC, SAR, etc.) also adversely affected.

Effects of Salts on Plants

As salts are enriches in soil, the osmotic pressure of soil water increases, and when this osmotic pressure is more than the osmotic pressure of roots, then plant is unable to absorb nutrients. Germination of seeds also reduces in salt affected soil because as radicle and plumule emerges, they injured with salts. Availability of some elements (Ca, Mg etc.) decreases and few elements (Na etc.) increases up to toxic levels. Some saline and alkali hazards on plants are given in Table 3.

Srivastava and Srivastava (1993) and Dalal (2001) reported that organic carbon decreases with increasing soil pH. Organic carbon and nitrogen have a positive correlation that's why as soil pH decreases nitrogen is also decreases (Fig. 1), resulting which soil become less fertile or unfertile. Chaney *et al.* (1992) reported that as HCO_3^- increases in the soil, soil pH increases and create buffer, which reduces the availability of Fe III and roots secrets phenolic substances, resulting which synthesis of CO_2 and organic acid is increases and whatever amount of Fe III is reached in roots a part of that amalgamated with organic acid and rest go to shoot this Fe III is uneven distributed in leaves that's why chlorophyll is formed in patches, so yield of plant is decreased (Fig. 2).

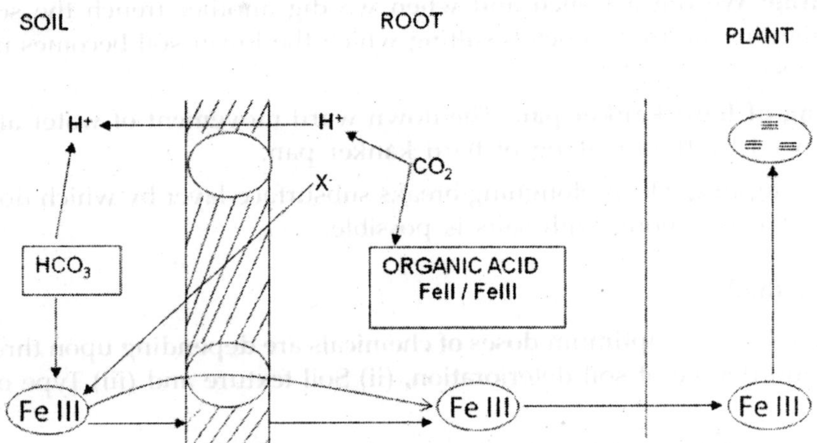

Figure 2: Effects of bicarbonate on chlorophyll formation in plants.

Table 3

Saline and alkali hazards on plants in relation to ESP and EC

(*Source*: Brady, 1996).

ESP	EC(dSM^{-1})	Saline / alkali hazard
< 15	<2	None to slight
15 – 30	2 – 4	Slight to moderate
30 – 50	4 – 8	Moderate to high
50 – 70	8 – 16	High to very high
70	> 16	Extremely high

Methods of Reclamation

Reclamation methods are mainly divided in to three viz. Physical, chemical and biological. Physical methods are crude and time consuming but these are most suitable for the reclamation of saline soils while chemical methods give quick response and suitable for reclamation of alkali soils. Biological methods are ecofriendly and reclaim both type of soils but they are time consuming.

Physical Methods

The following methods are found suitable to reclaim these problematic soils.

Scraping of salts: We scrape the salts and dumped them elsewhere.

Leaching of the salts: In this method field filled with water and salts are leached with water.

Drainage: In drainage huge amount of water leave in the field for a short period after that drained out in which salts are desolved.

Trenching: We dig a trench and when we dig another trench the soil of this trench is filled in the first trench resulting which the lower soil becomes upper soil and vice versa.

Breaking of hard kanker pan: The down ward movement of water along with salts is allowed by the breaking of hard kanker pan.

Deep ploughing: Deep ploughing breaks subsurface layer by which down ward movement of water along with salts is possible.

Chemical Methods

Optimum dose: The optimum doses of chemicals are depending upon three factors (i) Nature and degree of soil deterioration, (ii) Soil texture and (iii) Type of crop to be grown.

Each chemical firstly formed calcium sulphate ($CaSO_4$), then this $CaSO_4$ react with sodium binded clay particles and replaced Na^+ with Ca^{++} and formed sodium sulphate (Na_2SO_4), which percolate in lower strata with water and / or drained out from the field through facilitating drainage.

Use of gypsum

$2NaHCO_3 + CaSO\text{-}_4 \qquad CaCO_3 + Na_2SO_4 + CO_2$

$Na_2CO_3 + CaSO\text{-}_4 \longrightarrow CaCO_3 + Na_2SO_4$

Na^+

$\boxed{CLAY} + CaSO_4 \longrightarrow Ca^{++} \boxed{CLAY} + 2Na_2SO_4$

Na^+

Use Of Pyrite

$FeS_2 + 2H_2O + 7O_2 \longrightarrow FeSO_4 + 2H_2SO_4$

Na^+

$\boxed{CLAY} + CaSO_4 \longrightarrow Ca^{++} \boxed{CLAY} + 2Na_2SO_4$

Na^+

Use of Sulphur

$2S + 3O_2 \longrightarrow SO_3$

$SO_3 + H_2O \longrightarrow H_2SO_4$

$CaCO_3 + H_2SO_4 \longrightarrow CaSO_4 + CO_2 + H_2O$

Na^+

$\boxed{CLAY} + CaSO_4 \longrightarrow Ca^{++} \boxed{CLAY} + 2Na_2SO_4$

Na^+

Use of Sulphuric Acid

$CaCO_3 + H_2SO_4 \longrightarrow CaSO_4 + CO_2 + H_2O$

$$\begin{array}{l} Na^+ \\ \quad \boxed{CLAY} + CaSO_4 \longrightarrow Ca^{++} \boxed{CLAY} + 2Na_2SO_4 \\ Na^+ \end{array}$$

Use of Calcium Polysulphide

$$CaS_5 + 8O_2 + 4H_2O \longrightarrow CaSO_4 + H_2SO4$$

$$CaCO_3 + H_2SO_4 \longrightarrow CaSO_4 + CO_2 + H_2O$$

$$\begin{array}{l} Na^+ \\ \quad \boxed{CLAY} + CaSO_4 \longrightarrow Ca^{++} \boxed{CLAY} + 2Na_2SO_4 \\ Na^+ \end{array}$$

Use of Ferrous Sulphate

$$FeSO_4 + H_2O \longrightarrow FeO + H_2SO_4$$

$$CaCO_3 + H_2SO_4 \longrightarrow CaSO_4 + CO_2 + H_2O$$

$$\begin{array}{l} Na^+ \\ \quad \boxed{CLAY} + CaSO_4 \longrightarrow Ca^{++} \boxed{CLAY} + 2Na_2SO_4 \\ Na^+ \end{array}$$

Abrol *et al.* (1998) studied the relationship between chemical requirement with soil pH and texture and reported that chemical requirement increases with increasing pH and more chemicals require for reclamation of heavy soils followed by medium and light soil (Fig 3). Math *et al.* (1999) conducted a experiment on gypsum application and concluded that gypsum application decreases pH, EC and ESP of soil (Table 4).

Abrol et al. (1998) reported that by the application of F Y M in combination of gypsum, the growth of *casuarina equisetifoliya* plant is maximum with 100% survibility, while by the applecation of gypsum alone survibility is 100% but growth is less and by the application of FYM alone and in control survibility is 82% and 60%, respectively (Fig. 4).

Table 4

Effect of Gypsum application on soil properties (Source: Math et al., 1999).

pH		EC (dSM^{-1})		ESP	
BGA	AGA	BGA	AGA	BGA	AGA
8.7	8.1	1.21	1.00	20.0	16.0
8.5	8.0	0.50	0.30	31.0	24.0
9.2	8.5	2.40	0.86	15.8	10.5
9.5	8.9	1.90	0.35	19.9	14.4
9.6	8.8	2.40	0.84	28.4	14.5

BGA: before gypsum application
AGA: after gypsum application

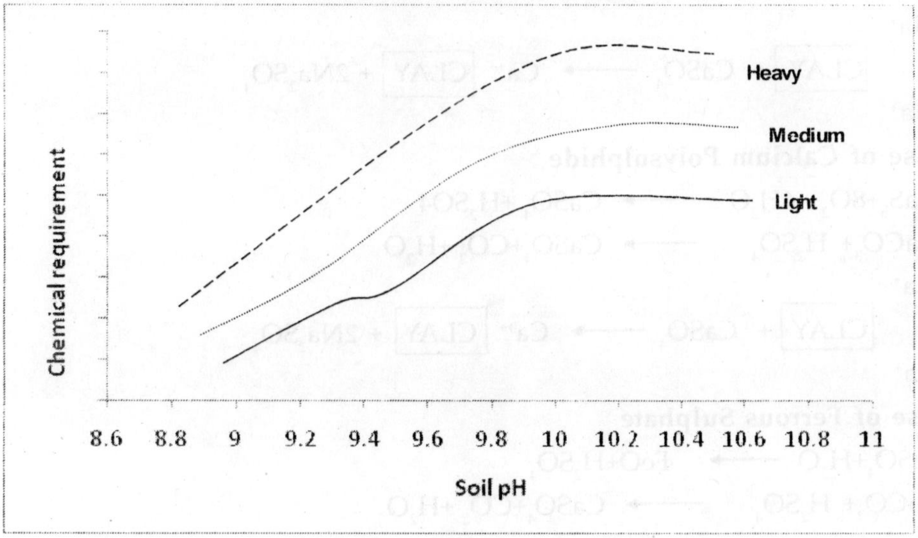

Figure 3: Relationship between chemical requirement with soil pH and texture (Source: Abrol et al., 1998).

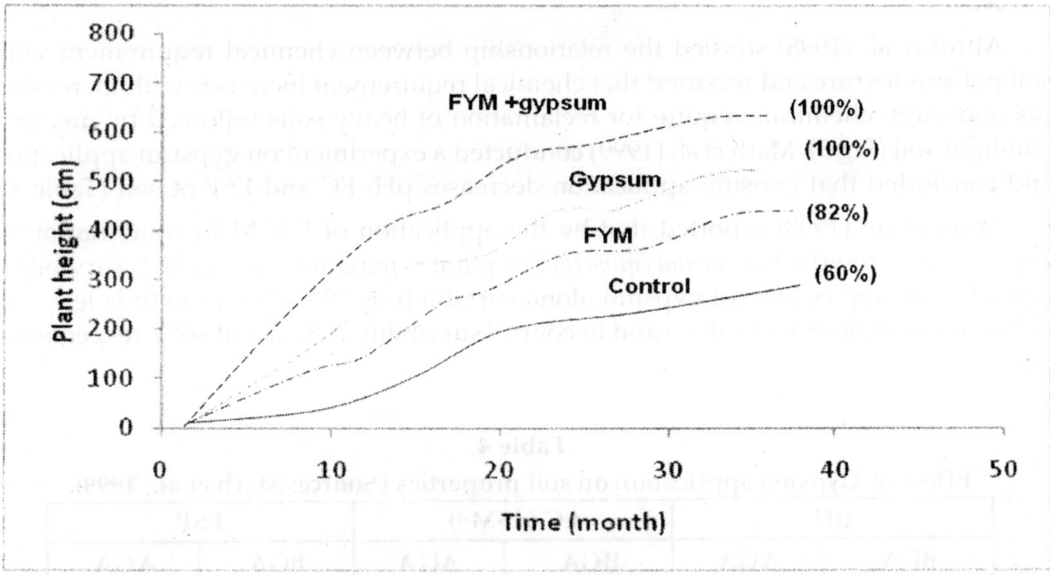

Figure 4: Effect of gypsum and FYM application on the growth of *Casurina equiselifolia* in sodic soils (values in parenthesis showing germination percentage) (Source: Abrol et al., 1998).

Biological Methods

Use of organic manure: After decomposition of organic manure like FYM compost etc humic and fulvic acids are formed which reclaims soil.

Use of green manure: By the use of green manure, chemical and physical properties of soil are improved and also increase microbial activity.

Use of molasses: By the addition of molasses organic acids are formed which reclaims the soil.

Mulching: Mulching directly reduces evaporation rate and also adds organic matter after decomposition.

Use of crop residue: Soil can also be reclaimed by use of crop residues directly or in the form of mulching, organic manure etc.

Growing of weeds: The roots of weeds like *Argemone maxicana* secrets chemicals, which reclaim soil.

Growing of salt tolerant plants: Soil can also be reclaimed, by growing of salt tolerant plants but it takes long time.

Tolerant Plants

Trees	Grasses and forage crops	Cereal Crops	Vegetable Crops	Fruit Crops
Babool	Alfa-alfa	Barley	Carrot	Ber
Jhau	Karnal grass	Rice	Lettuce	Aonla
Black Siris	Bermuda grass	Sorghum	Potato	Guava
Arjun	Para grass	Wheat	Onion	Phalsa
	Napier grass			Karonda
	Clover			

Kanthaliya *et al.* (1990) studied the effect of mulching at different depth of water table and reported that mulching reduces evaporation rate and EC of soil at each depth of water table (Table 5). Mongia and Chhabra (2000) analyzed the effect of agroforestry on chemical composition of alkali soil profile and found that by

Table 5

Effect of mulching on evaporation rate and soil EC (Source: Kanthaliya et al., 1990).

Depth of water table (cm)	Evaporation rate (mm/day)		EC (dSM^{-1})	
	UMS	MS	UMS	MS
25	4.0	1.1	10.9	9.3
50	4.0	1.4	10.0	9.0
75	2.9	1.3	7.0	5.1
100	1.7	1.1	5.2	2.7

UMS: Unmulched soil MS: Mulched soil

agroforestry pH and EC of soil reduces at each depth of soil profile (Table 6). Khanna (1994) conducted a experiment with various fruit crops model in sodic soils and after five year he reported that each fruit crops model reduces the pH, EC and ESP of soil at each soil depth but Guava and Ber model gave best result among them (Table 7).

Table 6

Effect of reclamation on chemical composition of alkali soil profile through agroforestry (*Source*: Mongia and Chhabra, 2000).

Depth of soil (cm)	pH		EC (dSM^{-1})	
	BFA	AFA	BFA	AFA
0-15	10.6	8.7	13.2	7.5
15-30	10.5	8.5	6.3	3.8
30-45	10.5	8.4	4.1	3.2
45-60	10.4	9.0	2.9	2.0
60-90	10.4	9.4	2.2	1.2
90-120	10.3	9.5	1.9	1.0
120-150	10.3	9.5	1.6	0.9
150-180	10.2	9.4	1.6	0.7

BFA: Before agroforestry
AFA: After agroforestry

Table 7

Change in properties of sodic soil under various fruit crop model after five year (*Source*: Khanna, 1994).

Model	Soil depth (cm)	pH	EC (dSM^{-1})	ESP
Control	0-15	10.4	18.3	76
	15-30	10.7	21.3	82
	30-45	10.6	24.2	84
Aonla+Guava	0-15	9.2	13.4	60
	15-30	10.0	15.5	66
	30-45	10.3	18.4	75
Aonla+Ber	0-15	9.5	12.0	57
	15-30	9.8	13.0	65
	30-45	10.2	15.0	73
Guava+Ber	0-15	9.0	8.3	47
	15-30	9.2	10.5	56
	30-45	9.6	14.0	65

Management

Proper management is essential after reclamation of such problematic soils therefore by adopting following techniques, soils can be managed successfully.

Ploughing: Ploughing should be avoided at same depth.

Puddling: Excessive puddling should be avoided.

Light and frequent irrigation: Irrigation should be applied lightly at frequent interval, which reduces evaporation.

Regular cultivation: Regular foliage cover achieved by adopting intensive crop rotation, so reducing the rate of evaporation.

Method of irrigation: It is beneficial to provide minimum water for irrigating the crops, for this drip and sprinkler irrigation method gave best result.

Drainage: Regular drainage is essential to avoid stagnation of excessive water in the field.

Fertilizers: Use of basic fertilizers like $CaCO_3$, basic slag etc should not be applied continuously.

Balba (1995) conducted an experiment using different bed shapes along with irrigation methods and reported that by using sloping bed, plants can be grown up to maximum salinity level i.e. 16 dSM^{-1} while by using rectangular bed, plants can be grown up to 8 dSM^{-1} in double row system and only up to 4dSM^{-1}in single row system irrespective of irrigation methods (Fig. 5).

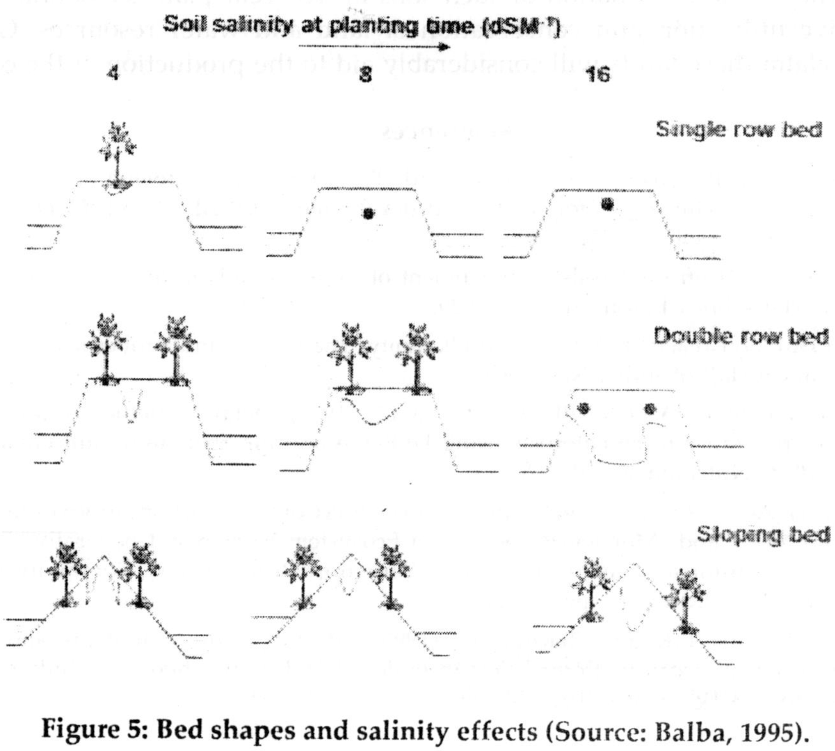

Figure 5: Bed shapes and salinity effects (Source: Balba, 1995).

Some constraints faced by farmers in adoption of technology:

1. The economy used by farmers in the use of fertilizers, amendment and land leveling is not cost effective.
2. Difficulty and inhibition on the part of farmers to get credit.
3. Fears that after reclamation, land would attract the provision of land ceiling act.
4. Long time taken by concerned department to supply electric connection to run tube well.
5. Lack of specialized extension staff earmarked to take up land reclamation in the state department of agriculture.

Conclusion

The salt affected soils present different kinds and intensities of problems. They are inherently rich in soil fertility and are high in production potential. However, these soils produce practicably nothing at present except some little grass. Though these soils may not produce bumper crops in the first year of reclamation but there will be gradual increase in production. For this specific ameliorative efforts requires like horizontal flushing and reclamation of soil by using physical, chemical and biological methods in combination and also need to develop appropriate policies and implement them. The reclamation of such soils in five-year plan is essential for the more effective utilization and conservation of land and water resources. Given a chance to reclaim these lands will considerably aid to the production of the country.

References

Abrol, I.P., Gupta, R.K., Joshi, P.K. and Prasad, R. (1998). Salt affected soils and their management. In: Technologies for waste land development. I.P. Abrol and V.V.D. Narayana (Eds.), ICAR, New Delhi, pp. 307-380.

Balba, A.M. (1995). Salt affected soils. Management of problem soils in arid ecosystems. CRC/ Lewis Publishers Boca Raton, FL, pp. 20-29.

Brady, N.C. (1996). Soil reaction: acidity and alkalinity. The nature and properties of soils, 10th edition. Printice Hall of India, New Delhi.

Chaney, R.L., Coulombe, B.A., Bell, PF. And Angle, J.S. (1992). Detailed method of screen dicot cultivars for resistance to Fe chlorosis using Fe-EDTA and bicarbonate in nutrient solution. Journal of Plant Nutrition 15: 2063-2083.

Dalal, R.C. (2001). Acidic soil ph, aluminium and iron affect organic carbon turnover in soil. In: Kirschbaum M.U.F.and Mueller, R. (eds.), Net Ecosystem Exchange. Cooperative Research Centre for Greenhouse Accounting, NEE Workshop Proceedings, Commonwealth of Australia 18 – 20 APRIL 2001, pp.111-115.

Kanthaliya, P.C., Chippa, B.R. and Sharma, S.L. (1990). Effect of water table depth and salinity of ground water on accumulation of salts in mulched and unmulched soils. Indian Journal of Soil Science, 23(1): 45-50.

Khanna, S.S. (1994). Management of sodic soils through tree plantation. Journal of the Indian Society of Soil Science, 42(3): 498-508.

Math, SKN., Rao, MSM, Padmaiah, M. and Patil, S.L. (1999). Effect of gypsum on soil properties and yield of rice in the loamy soils of semiarid tropics of Andhra Pradesh. Journal of Soil and Crops, 9(2): 145-150.

Mongia, A.D. and Chhabra, R. (2000). Silica and phosphate profiles of alkali soils following reclamation. Journal of the Society of Soil Science, 48(1): 33-37.

Srivastava, A.K. and Srivastava, O.P. (1993). Transformation and availability of nitrogen in salt affected soil. Indian Journal of Agricultural Chemistry, 26(2&3): 55-61.

2

Plant Resources as Insect Pest Management Agents

Deepa B. and Remadevi O.K.

Introduction

Plants are a rich source of primary and secondary metabolites which are intermediate products of enzyme catalyzed reactions that occur naturally within cells. Primary metabolites are directly involved in growth, development and reproduction. The secondary metabolites have insignificant role in primary physiological processes (Cooper and Johnson, 1984). Some of the secondary metabolites are the end products of aberrant biosynthetic pathways and others are excretory products. They have been attributed various ecological roles (Balandrin *et al.*, 1985). These principles perhaps are the defense strategies evolved by plants against the herbivore onslaught and, therefore, enhance the chance of survival of the species containing toxic constituents (McKey, 1984).

Commonly found secondary metabolites are acids, alkaloids, fatty acids, lactones, sterols, tannins, lignins, saponins, aldehydes, ester, glycosides, 'N' compounds, terpenoids, alcohols, flavones, hydrocarbons, phenols, polyacetylenes, unusual amino acids, sugars etc. These secondary metabolites exhibit different biological activities. The secondary metabolites of plants have been exploited by man since time immemorial in the form of medicines, cosmetics, dyes and pesticides. The plant world comprises a rich storehouse of biochemicals, which could be used as insecticides. Botanical insecticides, as these plant derived products are known, have been utilized by man since ancient times.

Botanical Insecticides in Use

The insecticidal properties of plants have long been recognized much before the advent of synthetic organic insecticides. In the middle of the 17th century, pyrethrum, nicotine, rotenone, isolated from *Chrysanthemum cinerariaefolium*, *Nicotiana* sp., *Derris elliptica* respectively, were recognized as effective insect control agents. Quassin, originally derived from the wood of *Quassia amara*, ryanodine derived from *Ryania speciosa* and sabadilla from *Schoenocaulony officinals* were the other products used as botanical pesticides in the earlier days (Benner, 1993 and Feinstein, 1952). In recent years, neem, *Azadirachta indica,* has gained popularity as potent botanical insecticide. A triterpenoid, azadirachtin was first isolated from neem seeds in 1968 and its structure was proposed in 1975. It is well known for its antifeedant, growth inhibitory and growth regulatory activities against various insect pests (Schmutterer, 1990; Koul, 1992, 1996). It has already been developed into various products and marketed. Other plants like *Vitex negunda, Pongamia pinnata, Madhuca indica* etc. are also commonly used for field pest management. Four furanoflavoines, viz., karanjin, pongapin, kanjone and pongaglabrone and a diketone pongamol, have been isolated from the seeds of *P. pinnata.* The oil and its active component karanjin possess insecticidal properties. *M. indica* contains 4.6% of a toxic and bitter saponin, viz., mowrin, which possess insecticidal properties (Jayaraj and Ignacimuthu, 2005). Other plants like Lantana, garlic, turmeric and essential oils from eucalyptus, mint and clove are used against various agricultural and stored product pests (Singh and Saratchandra, 2005; Isman, 2006). Botanical insecticides are used in the form of powders, volatile oils, non volatile oils, extracts and essential oils.

Important Plant Families with Insecticidal Activity

In India, more than 2400 plant species are known to possess insecticidal activity. They are distributed in 189 plant families (Singh, 2000). Some of the well known botanical insecticides like Nicotine, Pyrethrum, Derris, Hellebore, Anabasine belong to different families namely, Solanaceae, Compositae, Fabaceae, Liliaceae and Chenopodiaceae respectively. Anabasine is also found in Solanaceae family (Feinstein, 1952). Quassin is from Simarubaceae family and Ryanodine is from Flacourtiaceae family (Benner, 1993). The plant families Asteraceae, Fabaceae and Euphorbiaceae contain most of the insecticidal plant species reported (Singh and Saratchandra, 2005). In recent years, Meliaceae and Rutaceae have shown substantial promise in insect control, partly due to the presence of limonoids like azadirachtin or triterpenes, characteristic of the order Rutales (Champagne *et al.,* 1992). According to Jacobson (1989) the most promising botanical insecticides for use at the present time and in the future, are from the families Meliaceae, Rutaceae, Asteraceae, Malvaceae, Labiatae and Canellaceae. According to Singh *et al.* (1999) the number of plant species possessing insecticidal property in some of the families are: 39 in Apocynaceae, 70 in Asteraceae, 63 in Euphorbiaceae, 57 in Fabaceae, more than 500 in Meliaceae, 72 in Myrtaceae, 55 in Ranunculaceae, 27 in Rosaceae and 39 in Rutaceae.

Anti-insect Activity of Botanicals

The secondary metabolites of plants repel approaching insects, deter feeding and oviposition on plants, disrupt behaviour and physiology of insects in various ways and even prove toxic to different developmental stages of many insects. The plant extracts are being used as mortality agents, repellents, antifeedants, attractants, oviposition deterrents and sterility agents. Among the various natural products scanned for their bio-efficacy against insect pests, the indian neem tree, *A. indica,* the karanj tree, *P. pinnata* and madhuca tree, *M. indica* are endowed with an array of unique phytochemicals possessing diverse biological effects such as antifeedant, repellent, growth regulating and insecticidal (Schmutterer, 1990).

Plants with Insecticidal Properties

Knowledge of the plants, their active principles and their role in insect pest management is of paramount importance to utilize them commercially. A vast body of literature has been accumulated on insecticides occurring in plants during the last two-three decades in India and abroad.

Important Insecticidal Chemical Constituents from Plants

With the increasing number of plant species being tested for any type of biological activity, it becomes essential to trace out the specific chemicals responsible for such activities and isolation and identification of their active principles, which may further help in synthesis of the products. The biological activities exhibited by various botanical products are attributed to these specific chemicals of varying groups. Several classes of compounds with diverse active groups are alkaloids, flavonoids, terpenoids, glycosides, phenols, tannins etc. Now, a new class of insecticidal compounds is being introduced that offers a safe and effective alternative to conventional chemical insecticides. They are based on a family of compounds named "polyol esters", which are sugar esters. The best of them would provide relief to agricultural crops, household and garden plants (Bliss, 2005). Some important secondary compounds isolated, identified from botanicals and bioassayed against the insect pests are presented in Table 2.

Insecticides based on Plant Essential Oils

Plant essential oils consist of low molecular weight terpenes and phenols. These oils have played a prominent role as fragrances for the perfume industry and as flavouring agents in food industry. Recent research and development activities in several countries suggest that plant essential oils or their constituents may play yet another role as crop protection agents. Some of the essential oils with demonstrated toxic actions in insects are from clove, eucalyptus, lemon grass, penny royal, thyme, citrus, cinnamon etc. The major constituents of the above are eugenol, 1,8-cineole citronellal, citral, (+)pulegone, thymol, (+)limonene, cinnamaldehyde etc (Isman, 1999).

Conclusion

Most of the botanical insecticides, viz., neem, chinaberry, custard apple, plant essential oils etc. are characterized by low mammalian toxicity, reduced effects on other non target organisms and minimal environmental persistence though some plant compounds like nicotine and strychnine used for insect control are as toxic

Table 1
Some Important Plants with Insecticidal Property

Sl.No.	Botanical name	Common Name	Family	Part possessing insecticidal property	Source
1	*Albiziz lebbeck*	Kala siris, kala tendu	Mimosaceae	Seeds, leaves, petioles, pods	WOI(1985)
2	*Allium sativum*	Garlic	Liliaceae	Garlic juice	Prowse *et al.*, 2000
3	*Anacardium occidentale*	Cashew nut	Anacardiaceae	Cashewnut shell	Watt Breyer and Brandwij, 1962
4	*Apium graveolens*	Celery	Umbelliferae	Seeds	Choochote *et al.*, 2004
5	*Argemone mexicana*	Mexican prickly poppy	Papaveraceae	Leaf	Malarvannan *et al*, 2008
6	*Calophyllum inophyllum*	Sultanachampa	Guttiferae	Leaf	WOI(1985); Gupta *et al.*, 1988
7	*Calotropis gigantea*	Milk weed	Asclepiadaceae	Plant latex	WOI 1988
8	*Cassia sophera*	Andadasi,Senna sophera	Leguminosae	Leaf	Kestenholz *et al.*, 2007
9	*Citrus aurantium*	Orange	Rutaceae	Fruit	Siskos *et al.*, 2007
10	*Clerodendron inerme*	Kashmir bouquet	Verbenaceae	Leaf	Patil *et al.*, 2006
11	*Clerodendron siphonanthus C. inerme (=C. indicum)*	Bhrangi Chingari Lan jai	Verbenaceae	Leaf	WOI, 1988; Roychoudry, 1994
12	*Decalepis hamiltonii*	Swallowroot	Asclepiadaceae	Roots	George *et al.*,1999
13	*Derris ferruginea*	Tuba root	Fabaceae	Root	Nayar *et al.*, 1979
14	*Dodonaea angustifolia*	Dodonaea	Sapindaceae	Leaf	Subashini *et al.*, 2004
15	*Giliricidia sepium*	Madre tree	Papilionaceae	Leaf and Bark	Kayode, 2006
16	*Luffa acutangula*	Chinese okra	Cucurbitaceae	Leaf	Mullai and Jebanesan, 2007

17	*Madhuca indica (=Bassia latifolia)*	Mahua tree	Sapotaceae	Seed	Krishnaiah and Kalode, 1990; Jothi *et al.*, 1990; Sangappa, 1977; Gupta *et al.*, 1988
18	*Melia azaderach*	Chinaberry	Meliaceae	Stems, roots, leaf, fruit, bark	Kumar *et al.*, 1990;Mahla *et al.*, 2002; Defago *et al.*, 2006
19	*Melia toosendan*	Chinaberry	Meliaceae	Bark and seed kernel	Feng, 1995
20	*Ocimum gratissimum*	Wild basil	Lamiaceae	Leaf, shoot and inflorescence	Subramanian, 1999
21	*Polygonam orientale*	Pepper wort	Polygonaceae	Leaf and stem	Deka and Singh, 2005
22	*Ricinus communis*	Castor	Euphorbiaceae	Leaf	Quisumbing, 1951
23	*Shorea robusta*	Sal	Dipterocarpaceae	Seed	WOI, 1988; Ketkar, 1987
24	*Tagetes erecta*	African marigold	Asteraceae	Flower and leaf	Usha Rani *et al.*, 2007
25	*Terminalia paniculata and T. bellirica*	Bahera	Combretaceae	Seeds	Reddy and Urs, 1989
26	*Thevetia neriifolia*	Kaneer sufaid	Apocynaceae	Leaves and seeds	Bai and Koshy, 2004
27	*Vateria indica*	Indian copal tree	Dipterocarpaceae	Seeds	Gupta *et al.*, 1988; Reddy and Urs, 1992
28	*Vitex mollis*	Sinduas	Verbenaceae	Leaves	Lopez *et al.*,2007

and hazardous as many synthetic insecticides. Botanical insecticides are relatively specific in their mode of action and easy to process and use. They are less expensive, easily available and can be produced by farmers themselves. The real benefits of botanical insecticides can be best realized in developing countries, where farmers may not be able to afford synthetic insecticides and where traditional use of plants and plant derivatives for crop protection is long established. There is a growing market all over the world for organically produced fruits, vegetables, spices, beverages, grains, cotton, etc. In the context of agricultural pest management, botanical insecticides are best suited for use in organic food production.

Table 2
Insecticidal compounds from different plants

Sl. No.	Compound	Name of the plant	Family	Insect	Mode of activity	Reference
1	(2E)-1-(2-hydroxyphenyl) pent-2-en-1-one and 1-((3-hydroxy-5,5-dimethylcyclohex-3-en-1-yl)oxy)hexane-3-one	*Hyptis suaveolens*	Lamiaceae	*Helicoverpa armigera* and *Spodoptera litura*	Larvicidal, antifeedant, ovicidal, oviposition deterrent	Raja *et al.*, 2005
2	6β-HydroxygeduninNimbocinol, azadiradione and salanin	*Azadirachta indica*	Meliaceae	*Helicoverpa armigera* and *Spodoptera litura*	Growth inhibitory	Koul *et al.*, 2003; Koul, 2004
3	Acetogenin	*Asimia tribola*	Annonaceae	Some Lepidopteran insects	Insecticidal	Rupprecht *et al.*, 1990
4	Azadirachtin	*Azadirachta indica*	Meliaceae	*Otiorhynchus sulcatus*	Growth inhibitory	Cowles, 2004
5	Azadirachtin	*Azadirachta indica*	Meliaceae	*Schistocerca gregaria, Pieris brassicae* and *Dysdercus sp.*	Antifeedant and growth inhibitory	Gill and Lewis, 1971
6	Azadirachtin	*Azadirachta indica*	Meliaceae	*Ostrinia furnacalis*	Growth inhibitor, insecticidal	Chiu *et al.*, 1984
7	Azadirachtin	*Azadirachta indica*	Meliaceae	*Dysdercus koenigii*	Growth inhibitor, insecticidal	Koul, 1984
8	Azadirachtin	*Azadirachta indica*	Meliaceae	*Ostrinia furnacalis*	Insecticidal	Arnason *et al.*, 1985
9	Capillin	*Artemisia nilagirica*	Asteraceae	*Culex pipiens quinquefasciatus*	Larvicidal	Banerji *et al.*,1990
10	Cefrelene	*Cedrela toona*	Meliaceae	*S.litura*	Antifeedant	Kubo and Kloke, 1982
11	Chromones and Flavones	*Sculletaria baicarensis*	Rutaceae	*Spodoptera litura*	Antifeedant	Morimoto *et al.*, 2003
12	Coumaran and quinines	*Cyperus spp.*	Cyperaceae	*Spodoptera litura*	Antifeedant	Morimoto *et al.*, 1999
13	Diterpenoid	*Scutellaria rubicunda*	Lamiaceae	*Spodoptera litura* and *Helicoverpa armigera*	Antifeedant	Bruno *et al.*, 1999

14	Eugenol	*Eugenia ceromatien*	Myrtaceae	*Stored grain pests*	Larvicidal	Morgan and Mandava, 1987
15	Excelsin	*Ailanthus excelsa*	Simaroubaceae	*Spilarctia oblique*	Insecticidal and growth inhibitory	Prajapati *et al.*, 1999
16	Flavaglines	*Aglaia species*	Meliaceae	*Spodoptera litura*	Insecticidal	Greger *et al.*, 2001
17	Heliotropine	*Heliotropium peruvianum*	Boraginaceae	Body louse	Larvicidal	Louis Feinstein, 1952
18	Lactone glycoside	*Cleistanthus collinus*	Euphorbiaceae	*Spodoptera litura*	Insecticidal	Arivudainambi *et al.*, 2006
19	Limonoids	*Azadirachta indica*	Meliaceae	*Helicoverpa armigera*	Growth inhibitory and antifeedant	Murugan *et al.*, 1998
20	Limonoids	*Citrus reticulata*	Rutaceae	*Culex quinquefasciatus*	Moult inhibition	Jayaprakasha *et al.*, 1997
21	Matrine and oxymatrine	*Sophora flavescens*	Leguminosae	Termites	Insecticidal and antifeedant	Mao and Henderson, 2007
22	Meliartenin	*Melia azedarach*	Meliaceae	*Epilachna paenulata*	Insecticidal and antifeedant	Carpinella *et al.*, 2003
23	Meliocarpins	*Melia azedarach*	Meliaceae	*Spodoptera litura*	Insecticidal	Bohnenstengel *et al.*, 1999
24	Neo-clerodane diterpenes	*Clerodendron inerme*	Verbenaceae	Housefly	Antifeedant and growth inhibitory	Rao *et al.*, 1993
25	Neo-clerodanes	*Teucrium tomentosum*	Labiatae	*Spodoptera litura* and *Plutella xylostella*	Antifeedant	Krishnakumari *et al.*, 2003
26	Neo-clerodane diterpenoids	*Teucrium fruticans*	Labiatae	*Spodoptera litura*	Antifeedant	Bruno *et al.*, 1999
27	Nimbidinin	*Azadirachta indica*	Meliaceae	*Locusta migratoria*	Repellent	Pradhan *et al.*, 1963
28	Pyrazole compounds	*Cucumis species*	Cucurbitaceae	*Aphis gossypii*	Growth inhibitory and insecticidal	Chen *et al.*,1999
29	Pyrido(1,2-a)azepine alkaloid	*Stemona species*	Stemonaceae	*Spodoptera litura*	Insecticidal	Kaltenegger *et al.*,2003
30	Rhodojaponin	*Rhododendron molle*	Ericaceae	*S.frugipurda Smith H. virescens fab. Leptinotarsa decemlineata Say. Gramium Rondani.*	Antifeedant	Hu *et al.*, 1993
31	Rocaglamide	*Aglaia elaeagnoidea/Aglaia odorata*	Meliaceae	*Helicoverpa armigera/ Spodoptera litura*	Larvicidal	Koul, 2004; Nugroho *et al.*,1999

32	Salannin	*Azadirachta indica*	Meliaceae	*Epilachna varivestis*	Growth inhibitor and antifeedant	Schmutterer, 1984
33	Scabrin	*Heliopsis scabra*	Compositae	Housefly	Larvicidal	Louis Feinstein, 1952
34	Sesquiterpenes	*Chloroxylon swietenia*	Rutaceae	*Helicoverpa armigera*	Insecticidal and antifeedant	Ravi Kiran *et al.*, 2007
35	Thionemone	*Azadirachta indica*	Meliaceae	*Dysdercus koenigii*	Ovicidal	Agrawal, 1993
36	Toosendanin	*Melia azedarach*	Meliaceae	*Helicoverpa armigera*	Antifeedant and growth inhibitory	Koul *et al.*, 2002
37	Wilfordine	*Tripterygium wilfordii*	Celastraceae	DBM, Cockroach	Larvicidal	Louis Feinstein, 1952
38	Hildecarpin & Tephrosin	*Tephrosia hildeibrandtii*	Fabaceae	*Maruca testularis*	Larvicidal	Hassanali & Lwande, 1989

Table 3

Major insecticidal essential oils from different plants

S. No.	Plant from which essential oil is extracted	Family	Insect	Reference
1.	*Thuja occidentalis*	Cupressaceae	*C. maculates*	Keita *et al.* (2001)
2.	*Ocimum sanctum*	Lamiaceae	*Apis mellifera*	Gill, 2002
3.	*Pimpinella anisum*	Apiaceae	*Culex pipiens*	Erler *et al.* (2006)
4.	*Eucalyptus sp.*	Myrtaceae	*Culex pipiens*	Erler *et al.* (2006)
5.	*Mentha sp.*	Lamiaceae	*Culex pipiens*	Erler *et al.* (2006)
6.	*Ocimum basilicum*	Lamiaceae	*Culex pipiens*	Erler *et al.* (2006)
7.	*Malosma laurina*	Lauraceae	*Culex pipiens*	Erler *et al.* (2006)
8.	*Cymbopogon schoenanthus*	Poaceae	*C. maculates*	Ketoh *et al.*, 2006
9.	*Artemisia vulgaris*	Asteraceae	*T. castaneum*	Wang, 2007
10.	*Khaya senegalensis*	Meliaceae	*C. maculates*	Bamaiyi *et al.*, 2006
11.	*Chloroxylon swientia*	Rutaceae	*H. armigera*	Ravi Kiran, 2007.
12.	Citrus sinensis	Rutaceae	Termites	Chang *et al.*, 2001; Zhu *et al.*, 2001 and Raina *et al.*, 2007.
13.	Chrysopogon zizanioides	Poaceae	Termites	Chang *et al.*, 2001; Zhu *et al.*, 2001 and Raina *et al.*, 2007.
14.	Eucalyptus sp.	Myrtaceae	Termites	Chang *et al.*, 2001; Zhu *et al.*, 2001 and Raina *et al.*, 2007.
15.	*Syzygium aromaticum*	Myrtaceae	Termites	Chang *et al.*, 2001; Zhu *et al.*, 2001 and Raina *et al.*, 2007.
16.	Cedar wood sp.	Pinaceae	Termites	Chang *et al.*, 2001; Zhu *et al.*, 2001 and Raina *et al.*, 2007.
17.	*Cymbopogon sp.*	Poaceae	Termites	Chang *et al.*, 2001; Zhu *et al.*, 2001 and Raina *et al.*, 2007.
18.	*Taiwania cryptomerioides*	Cupressaceae	Termites	Chang *et al.*, 2001; Zhu *et al.*, 2001 and Raina *et al.*, 2007.

References

Agrawal, I.L. (1993). Ovicidal activity of four plant extracts on *Dysdercus koenigii* Fabr. (Heteroptera: Pyrrhocoridae). *Indian J. Ent.* 55: 440-443.

Arivudainambi, S., Selvamuthukumaran, T. and Baskaran, P. (2006). Lactone glycoside in *Cleistanthus collinus* (Euphorbiaceae), an effective insecticidal principle. *Indian Journal of plant protection,* 34(2): 256-257

Arnason, J.T., Philogene, B.J.R., donskov, N., Hudson, N., McDougall, C., Forster, G., Mrnd, P., Gardner, D., Lambert, J., Morris, C and Nozzolilla, C. (1985). Antifeedant and insecticidal properties of azadirachtin to European corn borer, *Ostinia nubilalis. Entomologia Experimentalis et. Applicat.,* 38 (1): 29-34.

Bai, H and Koshy, G. (2004). Juvenomimetic activity of extracts of *Thevetia neriifolia* Juss. to *Dysdercus cingulatus* F. (Hemiptera: Pyrrhocoreidae). *Journal of tropical agriculture,* 42 (1-2): 45-47.

Balandrin, M. F., Klocke, J. A., Wurtele, E. S. and Bollinger, Wm. H. (1985). Natural plant chemicals: sources of industrial and medicinal materials. *Science,* 228:1154

Bamaiyi, L. Ndams, I. S., Toro, W. A. and Odekina, S. (2006). Effect of mahogany *Khaya senegalensis* seed oil in the control of *Callosobruchus maculates* on stored cowpea. *Plant Protect. Sci.,* 42:130-134.

Banerji, A., Luthria, D. L. and Kokate, S. D. (1990). Toxicity of capillin, the insecticidal principle of *Artemisia nilagirica* Clarke. *Indian Journal of Experimental Biology,* 28: 588-589.

Benner, J. P. (1993). Pesticidal plants from higher plants. *Pestic Sci.,* 39: 95-102.

Bliss, R. M. (2005). Death by dessication. *Agricultural Research,* 8-9.

Bohnenstengel, F.I., Wray, V. Witte, R.P., Srivavastava and Prakash, P. (1999). Insecticidal meliacarpins (C-seco limonoids) from *Melia azedarach, Phytochemistry,* 50: 977-982.

Bruno, M., Vassallo, N. and Simmonds, M.S. (1999). A diterpenoid with antifeedant activity from *Scutellaria rubicunda. Phytochemistry,* 50: 973-976.

Carpinella, M.C., Defago, M.T, Valladares, G., Palacios, S.J. (2003). *Agri. Food Chem.* 51: 369.

Champagne, D. E., Koul, O., Isman, M. B., Scudder, G. G. E. and Towers, G.H.N. (1992). Biological activity of limonoids from the Rutales. *Phytochemistry,* 31:377-394.

Chang, S., Cheng, S., and Wang, S. (2001). Antitermitic activity of essential oils and components from Taiwania (*Taiwania cryptomerioides*). *Journal of Chemical Ecology,* 27(4): 717-724.

Chen, J., Rahbe, Y. and Delobe, B. (1999). Effects of pyrazole compounds from melon on the melon aphid *Aphis gossypii. Phytochemistry,* 50: 1117-1122.

Chiu. Shin Foon, Zhang, X., Liu, S. K. and Huang, D.P. (1984). Growth-disruptive effects of Azadirachtin on the larvae of Asiatic corn borer, *Ostrinia furnacalis* Guen. *Acta Entomologica sinica,* 27(3): 241-247.

Choochote, W., Benjawan, T., Duangta, K., Eumporn, R., Udom, C., Prosong, C., Jitpakdi, A ., Ponsgsri, T. Riyong, D. and Benjawan, P. (2004). Potential of crude seed extract of celery, *Apium gravelens* L., against the mosquito *Aedes aegypti* (L.) (Diptera: Culicidae). *Journal vector ecology* , 29(2): 340-346.

Cooper, M.R. and Johnson, A.W. (1984). Poisonous plants in Britain and their effects on Animals and Man. Reference Book, 161, Ministry of Agriculture, fisheries and food, London.

Cowles, R.S. (2004). Impact of azadirachtin on vine weevil (Coleoptera: Curculionidae) reproduction. *Applied and Forest Entomology,* 6: 291-294.

Defago, M, Valladares, G. Banchio, E., Carpinella, C., Palacios, S. (2006). Insecticide and antifeedant activity of different plant parts of *Melia azadarach* on *Xanthogaleruca luteola, Fitoterapia* 77: 500-505.

Deka, M. K. and Singh, K. (2005). Effect of aqueous plant extracts of *Clerodendron inerme* and *Polygonum orientale* on growth and development of tea mosquito bug (*Heliopeltis theivora* Waterhouse). *Indian Journal of Entomology,* 67(2):93-96.

Erler, F., Ulug, I., and Yalcinkaya, B. (2006). Repellent activity of five essential oils against *Culex pipiens. Fitoterapia,* 77: 491-494.

Feinstein, L. (1952). Insecticides from plants. Year Book of Agriculture. 222-228

Feng, R., Chen, W., and Isman, M.B. (1995). Synergism of malathion and inhibition of midgut esterase activity by an extract from *Melia toosendan. Pesticide Biochemistry and Physiology,* 53: 34-41.

George, J., Ravishankar, G.A., John Pereira and Divakar, S. (1999). Bioinsecticide from swallowroot (*Decalepis hamiltonii*) Wight & Arn protects food grains against insect infestation. *Current Science,* 77(4):501-503.

Gill, J.S. and Lewis, C.T. (1971). Systemic action of an insect feeding deterrent. *Nature* 233 : 402-403.

Gill, K., Malik, S.B., Kalidhar, R.C., Mishra and Malik, M.S. (2002). Honey bee repellent activity of leaf extract and essential oil of *Ocimum sanctum* and isolation of a new compound. *Journal of medicinal and aromatic plant sciences,* 24:951-956.

Gregar, H., Pacher, T., Brem, B., Bacher, M., Hofer, O. (2001). Insecticidal flavaglines and other compounds from Fiijan *Aglain* species, *Phytochemistry,* 57: 57-64.

Gupta, P.S., Vimala, V., Geervani, P. And Yadagir, B. (1988). Efficacy of vegetable oils as protectants of greengram stored in different jute bags. *J.Food Sci. Tech.,* 25 (4): 194-196.

Hassanali, A. and Lwande, W. (1989). Anti-pest secondary metabolites from African plants. In: Insecticides of plant origin. ACS Symp. Series 387 of III Chem.Congr. North America 195th National Meeting of *Am. Chem. Soc.* held at Toronto, Ontario, Canada, 5-11 June, 1988: 78-94.

Hu, J., Yang , J. and Chen, L. (1993). A preliminary study on the antifeedant and to xic properties of China berry, *Melia azadirach* Linn. Seed oil against major insect pest of rice. *Scientia agric. Sinica,* 5: 63-69.

Isman, M.B. (1999). Pesticides based on plant essential oils. *Pesticide outlook,* 68-72.

Isman, M.B. (2006). Botanical insecticides, deterrents and repellents in modern agriculture and an increasingly regulated world. *Annu. Rev. Entomol.,* 51: 45-66.

Jacobson, M. (1989). Botanical Pesticides : past, present and future. In *Insecticides of plant origin.* J. T. Arnason, B. J. R. Philogene and P. Morand (Eds.). *ACS Symp. Ser.* 387, American Chemical Society, Washington, D. C. 1-10p.

Jayaprakasha, G.K., Singh, R.P., Pereira and Sakariah, K.K. (1997). Limonoids from *Citrus reticulata* and their moult inhibiting activity in mosquito *Culex quinquefasciatus* larvae. *Phytochemistry,* 44(5): 843-846.

Jayaraj, S. and Ignacimuthu, S. (2005). Botanical pesticides for sustainable Crop Protection. In Sustainable Insect Pest Management. Ignacimthu, S. and Jayaraj, S. (Eds.), Narosa Publishing House, New Delhi, 161-175p.

Jyothi, D.B., Verghese, A. and Tandan, P.L. (1990). Evaluation of different plant oils and extracts against citrus aphid, *Toxoptera citricidus* (Kirkaldy). *Indian Journal of Plant Protection* , 18: 251-254.

Kaltenegger, E., Brem, B., Mereiter, K., Kalchhauser, H., Kahlig, Hanspeter, K., Hofer, O., Vajrodaya, S. and Greger, H. (2003). Insecticidal pyrido (1,2-a) azepine alkaloids and related derivatives from *Stemona* species. *Phytochemistry*, 63: 803-816.

Kayode, J. (2006). Pesticidal activity of the leaf and Bark extracts of *Giliricidia sepium* on *Callosobrucus maculates*, *Journal of sustainable forestry*, 22(3/4):57-61.

Keita, S.K., Vincent, C., Schmidt, J., Arnason, J. (2001). Insecticidal effects of *Thuja occidentalis* (Cupressaceae) essential oil on *C. maculatus* (Coleoptera: Bruchidae). *Canadian Journal of Plant Sciences*, 81(1): 173-177.

Kestenholz, C., Stevenson, P.C. and Belmain, S.R. (2007). Comparative study of field and laboratory evaluation of the ethnobotanical *Cassia sophera* L. (Leguminosae) for bioactivity against the storage pests *Callosobruchus maculates* (F.) (Coleoptera : Bruchidae) and *Sitophilus oryzae* (L.) (Coleoptera : Curculionidae). *Journal of Stored Products Research*, 43: 79-86.

Ketkar, C.M. (1987). Use of tree derived non-edible oils as surface protectants for stored legumes against *C. maculates* and *C. chinensis*. Natural pesticide from the neem tree (*Azadirachta indica* A. Juss). and other tropical plants. Proceedings of the third 3rd International Neem Conference, Nairobi, Kenya, 10-15 July, 1986 (Schmutterer, R and Ascher, K. R. S.): 535-542.

Ketoh, G.K., Koumaglo, H.K., Glitho, I.A., Huignard, J. (2006). Comparative effects of *Cymbopogon schoenanthus* essential oil and piperitone on *C. maculates* development. *Fitoterapia*, 506-510.

Koul, O. (1984). Azadirachtin II. Interaction with the reproduction behaviour of red cotton bugs. *Zeitschnift fur Angewandte Entomologei* 96(2): 221-223.

Koul, O. (1992). Neem allelochemicals and insect control. In: Allelopathy Basic and Applied aspects. Rizvi, S. J. H., Rizvi, R. (Ed.). London Chapman & Hall. 389-412.

Koul, O. (1996). Neem Research and Development: Present and future scenario. In: Supplement to cultivation and utilization of medicinal plants. Handa, S.S. and Koul, M. K. (Eds). National Institute of Science Communication, CSIR, New Delhi, 583-611.

Koul, O. (2004). Neem: a global perspective. In Koul ,O. & Wahab,S. (eds):N eem: today and in the new millennium . Kluwar Academic publishers, Dordrecht, Boston, London, 1-19p.

Koul, O., Multani, J.S., Singh, G. and Daniewsk, W. M. and Berlozecki, S. (2003). 6â-Hydroxygedunin from *Azadirachta indica*, its potentiation effects with some non-azadirachtin limonoids in neem against lepidopteran larvae. *J. Agric. Food Chem.*, 51 :2937-2942.

Koul, O., Multani, J.S., Gurmeet Singh and Seema Wahab. (2002). Bioefficacy toosendanin from *Melia dubia* (syn. *M. azedarach*) against gram pod-borer, *Helicoverpa armigera* (Hubner). *Current science*, 83 (11): 1387-1391.

Krishna kumarai, G.N., Aravind, S., Balachandran, J., Ganesh, Soundarya Devi, Rajan, S.S., Malathi, R. and Ravikumar (2003). Antifeedant neo-clerodanes from Teucrium tomentosum Heyne. (Labiatae), *Phytochemistry* 64 : 1119-1123.

Krishnaiah, N.V. and Kalode, M.B. (1990). Efficacy of selected botanicals against rice insect pests under green house and field conditions. *Indian J. Pl. Prot.*, 18: 197-205.

Kubo, I., Kloke, J.A., Miura, I. and Fukuyama, Y. (1982). Structure of ajugarin-IV. *J. Chem.Soc. chem. comm.* : 618p

Kumar, R., Ram singh, Meena Rani, Puja Suhag and Kalidhar, S.B. (1990). Biological efficacy of *Melia azadarach* stems against lepidopterous pests, *Journal of Medicinal and Aromatic plant science*, 21 : 1074-1078.

Lopez, V. R., Suarez, M. Z. F., Rodriguez, T. and Aranda, E. (2007). Insecticidal activity of *Vitex mollis*. *Fitoterapia*, 78: 37-39.

Louis Feinstein (1952). Insecticides from plants. Year Book of Agriculture. 222-228

Mahla, M., Ram Singh, Puja suhag, Bharti and Suraj, B. Kalidhar (2002). Bioevaluation Melia azedarach L. roots against *Helicoverpa armigera* and *Plutella xylostella, Journal of Non-timber products*, 9(1/2): 50-53.

Malarvannan, S., Senthil Kumar, S., Giridharan, R. and Sudha Nair (2008). Bioefficacy of *Argemone mexicana* against American boll worm, *Helicoverpa armigera* (Hubner) (Noctuidae: Lepidoptera). *Hexapoda*, 15(1): 49-55.

Mao, L. and Henderson, G. Antifeedant activity and acute and residual toxicity of alkaloids from *Sophora flavescens* (Leguminosae) against Formosan subterranean termites (Isoptera: Rhinotermitidae). *Journal of economic entomology*. 100(3): 866-870.

McKey, D. (1984). Interaction of the ant- plant *Leonardoxa africana* (Caesalpiniaceae) with its obligate inhabitants in a rainforest in Cameroon. *Biotropica*, 16:81.

Morgan, E.D. and Mandava, N.B. (1987). *Handbook of Natural Pesticides*. Vol. III. Insect growth regulators Part B, CRC Press, Boca Raton, Florida. 180p.

Morimoto, M., Tanimoto, K., Nakano, S., Ozaki, T., Nalano, A. and Komal, K. (2003). *J. Agric. Food. Chem.*, 51: 389-393.

Morimoto, M., Yuko Fuji, Komai, M. (1999). Antifeedant in cyperaceae: coumaran and quinines from Cyperus spp. *Phytochemistry*, 605-608.

Mullai, K. and Jebanesan, A. (2007). Larvicidal activity of leaf extract of *Luffa acutangual* Roxb. against *Anopheles stephensi* . *Insect environment*, 12(4):180-182.

Murugan, K., Jeyabalan, D., Senthil Kumar, N., Babu, R., Sivaramakrishnan, S. and Senthil Nathan, S. (1998). Antifeedant and growth inhibitory properties of neem limonoids against the cotton bollworm, *Helicoverpa armigera* (Hubner). *Insect Sci. Applic.*, 18(2): 157-162.

Nayar, K.K., Ananthakrishnan, T.N. and David, B.V. (1979). General and applied Entomology. Tata McGraw-Hill Publishing Company Ltd., New Delhi, p. 386.

Nugroho, B.W., Edrada, R.A., Wray, V., Witte, L., Bringmann, G., Gehling, M. and Proksch, P. (1999). An insecticidal rocaglamide derivative and related compounds from *Aglaia odorata* (Meliaceae). *Phytochemistry*, 51: 367-376.

Patil, P.B., Holihosur, S.N. and Kallapur, V.L. (2006). Efficacy of natural product, *Clerodendron inerme* against dengue mosquito vector *Aedes aegypti*. *Current Science*, 90(8):1064-1066.

Pradhan, S., Jotwani, M.G. and Rai, B.K. (1963). The repellent properties of some neem products. *Bull. Reg. Lab.* 1(2): 149-151.

Prajapathi, V., Tripathi, A.K., Jain, D.C., Sharma, S. and Khanuja, S.P.S. (1999). Toxic and growth inhibitory effects of excelsin isolated from *Ailanthus excelsa* against *Spilarctia oblique*. Biopesticides in insect pest management. S. Ignacimuthu, S.J. and Alok sen (eds.). Phoenix publishing House Pvt. Ltd. New Delhi, p. 128-134.

Prowse, G., Tamara, S. Galloway and Andrew, F. (2000). Insecticidal activity of garlic juice in two dipteran pests. *Agricultural and Forest Entomology* , 8: 1-10.

Quisumbing, E. (1951). Medicinal plants of the Phillipines. *Tech Bull. 16, Manila Bureau of Printing* 146p.

Raina, A., John Bland, Mark Doolittle, Alan Lax, Raj boopathy and Folkins (2007). Effect of Orange oil extract on the Formosan subterranean termite (Isoptera : Rhinotermitidae). *J. Econ. Entomol.*, 100(3) : 880-885.

Raja, N., Elumalai, M., Jayakumar, A., Jeyasankar, C. and Ignacimuthu (2005). Biological activity of different plant extracts against armyworm, *Spodoptera litura* (Fab.) (Lepidoptera: Noctuidae). *J.Ent. Res.*, 27(4): 281-292.

Rao, M. J.L., Pereira, J., and Gurudutt, K.N. (1993). Neo-clerodane diterpenes from *Clerodendron inerme, Phytochemistry*, 34: 2 572-574.

Ravi Kiran, S., Sita Devi, P. and Reddy, K.J. (2007). Bioactivity of essential oils and sesquiterpenes of *Chloroxylon swietenia* DC against *Helicoverpa armigera*. *Current Science*, 93(4): 544-48.

Reddy, G.V.P. and Urs, K.C.D. (1989). Antifeedant and repellent activity, of some indigenous plant extracts against the furniture beetle, *Sinoxylon sudanicum* Lesne (Coleoptera: Bostrychidae). In: Biodeterioration of cultural property (eds.) Agarwal, O. P. and Dhawan, S. 1989. Macmillan India Ltd. 173-185p.

Roychoudhury, N. (1994). Antifeedant and insecticidal activity of *Clerodendron* spp. against rice weevil. *Journal of Applied Zoological Research* 5(1): 13-16.

Rupprecht, J.K., Hui, Y.H, and Mclaughlin, J.L. (1990). Annonaceous acetogenins, a review. *J. Nat. Prod.* , 53: 237-278.

Sangappa, H.K. (1977). Effectiveness of oils and surface protectants against the bruchid, *Callosobruchus chinensis* Linn, infestation on red gram. *Mysore J. Agri, Sci.* 11: 391-397.

Schmutterer, H. (1990). Properties and potential of natural pesticides from the neem tree. *Ann. Rev. Ent.*, 35:271-297.

Schmutterer, H. and Ascher, K.R.S. (1984). Natural pesticides from the Neem Tree (*A. indica* A. Juss.) and other tropical plants. Proc. 2nd Int. Neem conf., Rauischhlzhausen, West Germany, 25-28 May 1983, GTZ, Eschborn (FRG), 587p.

Singh, R.N. and Saratchandra, B. (2005). The Development of Botanical Products with Special Reference to Seri- Ecosystem. *Caspian Journal of Environmental Sciences*, 3 (1): 1-8

Singh, R.P. (2000). Botanicals in pest management: An ecological prospective. In: G. S. Dhaliwal and B. Singh (eds.). Pesticides and Environment. Commonwealth Publishers, New Delhi, 279-343p.

Singh, R.P., Singh, S. and Wahab, S. (1999). Biodiversity and importance of botanical pesticides. In: G. S. Dhaliwal, R. Arora and A. K. Dhawan (eds.). Emerging trends in Sustainable Agriculture. Commonwealth Publishers, New Delhi, 282-299p.

Siskos, E.P., Konstantopoulou, M.A., Mazomenos, B.E., Jervis, M. (2007). Insecticidal activity of *Citrus aurantium* fruit, leaf, and shoot extracts against adult Olive fruit flies (Diptera: Tephritidae). *Journal of economic entomology*, 100(4) : 1215-1220

Subashini, H.D., Malarvannan, S. and Pillai, R.R. (2004). *Dodonaea angustifolia* – a potential biopesticide against *Helicoverpa armigera*. *Current Science*, 86(1): 26-28.

Subramaniam, B. (1999). Underexploited insecticidal plants of potential agricultural value. Biopesticides in insect pest management. S. Ignacimuthu, S.J. and Alok sen (eds.). Phoenix publishing House Pvt. Ltd. New Delhi. 71-83p.

Usha Rani, P. and Udaya lakshmi, S. (2007). *Tagetes erecta* (L.) essential oil, floral and foliar extracts as potential grain protectants against three species of stored product pests. *Uttar Pradesh J. Zool.* 27(2): 159-169.

Wang, Y.N., Shi, G.L., Liu, L. Yu, T.Q., Clarke, S.R. J.H. (2007). Toxicity of plant essential oils and their components against *Lycoriellata ingénue* (Diptera: Sciaridae). *Journal of economic entomology*, 100(6): 1298-1303.

Watt, J.M. and Breyer-Brandwijkm, M.C. (1962). Medicinal and poisonous plants of southern and Eastern Africa. E. & S. Livingstone Ltd., Edinburgh 213p.

WOI (The Wealth of India). (1985). A Dictionary of Indian Raw materials and industrial products. Raw materials, Vol I: A Publication and information Directorate, CSIR, New Delhi, pp. 302-305.

WOI (The Wealth of India). (1988). A Dictionary of Indian Raw materials and industrial products. Raw materials, Vol 2B: A Publication and information Directorate, CSIR, New Delhi.

Zhu, B. C.R., Henderson, G., Feng Chen, Fei, H. and Laine, R.A. (2001). Evaluation of vetiver oil and seven insect-active essential oils against the Formosan subterranean termite. *Journal of Chemical Ecology*, 27(8): 1617-1625..

3

Pulp and Paper Mill Wastewater and Its Various Treatment Techniques

N. Kumara Swamy and Pratibha Singh

Abstract

Pulp and paper mills use a variety of bio-resources to produce paper and generate a variety of pollutants depending on the type of the production processes adopted. This chapter focuses on the paper production techniques, pollutants generation and the performance of different available wastewater treatment processes related to paper industry. The different treatment processes such as coagulation, chemical oxidation, ozonation, aerobic and anaerobic treatment for reduction of color and other pollutants from paper industry wastewater are reviewed and discussed. Further, it is concluded that the chemical coagulation, oxidation and ozonation are effective in reduction of color from paper industry wastewater. The chlorinated phenolic compounds and adsorbable organic halides (AOX) are efficiently reduced by adsorption, ozonation and membrane filtration techniques. The combinations of anaerobic and aerobic treatment processes are found to be efficient in the removal of soluble biodegradable organic pollutants.

(**Key Words**: Pulping, Bleaching, Wastewater)

INTRODUCTION

Pulp and Paper Industry History and Status

The technique of paper-making is believed to have originated in China, from where it spread to the rest of the world including India. The first handmade paper mill in India was set-up in Kashmir in the 14th century. Much later, in 1832, the first mechanized paper mill based on jute and grass was established in Serampore, West Bengal. After a slow start, the growth was encouraged by the Bamboo Protection Act (1925) and Indian Finance Act (1931). More than 10 paper mills were commissioned in this period and by 1931, the production capacity had reached 45,600 tonnes.

India had 17 paper mills at the time of independence with an installed capacity of less than 0.14 million TPA (tonnes per annum). At present, there are an estimated

525 pulp and paper mills with a total installed capacity of around 6.25 million TPA (Singh 2004) with a capacity utilization of about 67%. The aggregate installed capacity by 2010 for paper and paperboard is expected to reach 8.3 million tonnes and 1.5 million tonnes for newsprint. The per capita consumption of paper in India is one of the lowest in the world at 5 kg. The present level of paper consumption in the country is 4.2 million tones. According to an estimation, the demand for paper and allied products is expected to cross 10 million tonnes by 2015 (Singh 2004).

Raw Materials and Waste Water

Pulp and paper are manufactured from raw materials containing cellulose fibers, generally wood, recycled paper, and non-wood raw materials such as bagasse, cereal straw, bamboo, reeds, esparto grass, jute, flax, and sisal. The manufacturing process uses a large amount of fresh water, most of which is thrown out as wastewater. Pulp and paper industry is one of the largest and most notorious sources of industrial pollution. The Ministry of Environment and Forest, Govt. of India, has categorized the pulp and paper industry as one of the twenty most polluting industries (Singh 2004).

The wastewater of from pulp and paper industries contains stray wood chips, bits of bark, cellulose fibres, dissolved ligneous material (30-45%), saccharinic acid (25-35%), formic acid and acetic acid (10%) and extractives (3-5%). The pulping and bleaching effluents are characterized by parameters unique to these wastes such as colour and organic halides (AOX). The residual lignin present in wood fibre is major coloring material, and also reacts with chlorine molecules and forms organochlorine compounds in the effluent. About 20% of the organically bound chlorine found in bleaching effluent corresponds to relatively low molecular mass (M.W 4000) products (Rintala et al. 1994).

The untreated wastewater from pulp and paper mills is generally discharged into water bodies and which cause damage to the water quality. The effluent imparts brown colour to water which is detectable over long distances. The effluents have high biological and chemical oxygen demands (BOD and COD), lignin compounds and their derivatives. The undiluted effluents are toxic to aquatic organisms and exhibit a strong mutagenic effect. Further more, some compounds in the effluents are resistant to biodegradation and can bioaccumulate in the aquatic food chain (Kumara Swamy 2008).

CLASSIFICATION OF PAPER MILLS

The categorization of pulp and paper mills is based on the raw material used, plant size, and end products manufactured.

Based on Raw Materials

Based on the raw materials used, the paper mills are categorized as follows:

Wood/forest Based Mills

These mills uses imported pulp as well as indigenous hardwood pulp from bamboo, eucalyptus, etc. The Indian paper industries, on an average, consume about 3%–4% of the total wood in India.

Agro-residue based Mills

These mills uses agricultural residues such as rice straw, wheat, sarkanda grass, bagasse, jute, etc, as raw materials. The use of agricultural residue by these mills has grown since the early 1970s partly due to the dwindling bamboo resources and partly due to the government's industrial policy encouraging investments in agro-based paper production. However, seasonal availability, transportation costs, and investments in pollution control equipment are seen as limiting factors.

Wastepaper-based Mills

These mills use imported and indigenous wastepaper, corrugated waste paper, kraft paper, and waste cuttings as raw materials. The recovery of wastepaper by these mills for paper production has increased from 65 000 tonnes in 1995 to 850 000 tonnes in 2000. However, the 20% rate of recovery is still one of the lowest, internationally. The statistics of raw material consumption by various mills is shown in Figure-1.

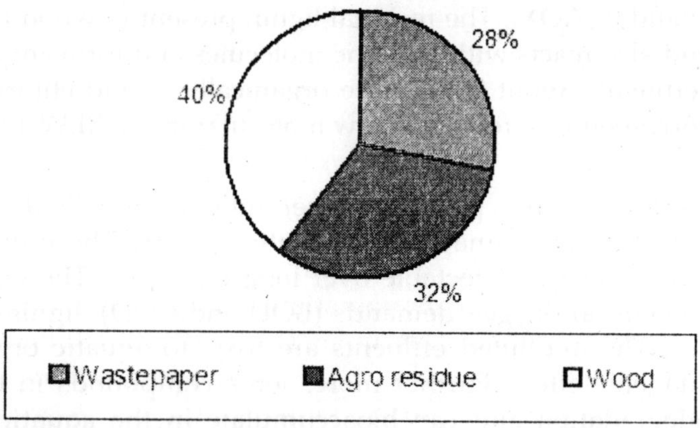

Figure 1: Share of paper production based on raw material (Singh 2004)

Based on Mill/Plant Capacity

The pulp and paper units can be broadly classified into small, medium and large mills based on the installed capacity.

Small Mills

These mills have production capacity less than 10,000 TPA capacity. These are further divided into mills of less than 2000 TPA capacity, 2000–5000 TPA capacity and 5000–10000 TPA capacities.

Medium Mills

These mills have production capacity of 10000 to 20000 TPA.

Large Mills

These mills have a capacity greater than 20000 TPA.

Table-1
Category-wise paper mills in India (In paper 2000)

Category	Capacity Range	Number of units	Capacity (TPA)
Small	Up to 10000	299	1290382
	<2000	69	75522
	2000–5000	107	296980
	5000–10000	123	917880
Medium	10000–20000	116	1669460
Large	>20000	70	3893048

The average size of a paper mill in India is 10,400 TPA, compared with 85, 000 TPA in Asia and about 3, 00, 000 TPA in Europe and North America. Of the total mills with a combined production capacity of around 68, 52, 890 TPA about 60% (299) are small mills. The small- and medium-sized mills together constitute 85% of the units but contribute to just 40% of the total production. The small units suffer from high production costs, uneconomic operation, low quality and negative impacts on environment. The large paper mills in India are much smaller in size when compared to mills in the Europian Union (EU). They are also relatively less advanced in terms of technology and machinery used. However, they are more efficient when compared to the smaller Indian mills.

Based on End Products

This classification uses **the type of end-product produced as the criterion.**

Pulp Mills

These are the mills which produce only pulp. The pulp is used for further processing.

Paper Mills

These mills produce only paper out of purchased pulp and/or recycled wastepaper.

Integrated Pulp and Paper Mills

These mills combine both the pulp and paper production and these are the most common type of units.

The end products produced by paper mills includes paper and paperboard viz. duplex board and paper, culturable paper viz. writing and printing paper, speciality paper such as security paper, tissue paper, cigarette paper. The newsprint culturable paper constitutes the biggest share with 41% production, followed by kraft paper (27%), paper board (17%), newsprint (12%) and speciality paper (3%).

Paper Production Process

The pulp and paper industry converts fibrous lignocellulosic raw material into pulp and paper. The typical pulp and paper manufacturing processes involved and different waste waters generated are shown (Figure-2) and discussed below.

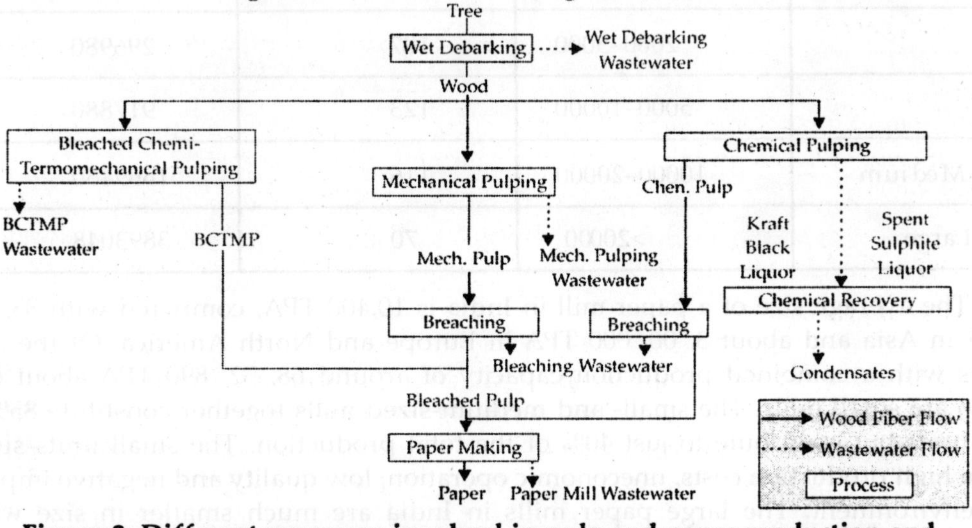

Figure-2: Different processes involved in pulp and paper production and corresponding waste waters generated (Jaakko and Puhakka, 1993).

Selection of Raw Material

The first step in pulp and paper production involves selection of suitable raw material keeping in mind the nature of end product desired. The primary material required in the manufacture of paper is cellulose fiber which is a biopolymer abundantly found in plant materials along with hemicellulose and lignin. The plant wood consists of cellulose (40-45%), hemicellulose (20-30%) and extractives (2-5%) (Singh 2004).

Wood cum forest based mills generally uses imported pulp and indigenous hardwood pulp from bamboo, eucalyptus, etc. The agro-residue based mills uses agricultural waste residues such as rice straw, wheat, sarkanda grass, bagasse, jute/ rags. Wastepaper-based mills employ imported and indigenous wastepaper, corrugated waste paper, kraft paper, and waste cuttings as the raw materials.

Debarking Process

Debarking process involves removing the bark and muddy materials from the surface of raw wood. In debarking process, the water is used to thaw the frozen logs, remove the bark and wash the logs after debarking. The wastewater generated contributes to about 15-20% towards COD, BOD and suspended solid. It is found to contain tannins, monomeric phenols, simple carbohydrates and resin compounds (Field et al. 1993).

Pulping Process

Pulping is a process performed to separate cellulose fibers from rest of the wood. The quality of pulp produced determines the quality of paper. The pulp with higher cellulose content and longer fibers yields a better quality paper. The use of hard woods reportedly produces higher proportion of cellulose than the soft wood (Jukka A Rintala et al. 1994). The pulping is performed using chemicals or mechanical or other methods.

Mechanical Pulping

Mechanical pulping involves the application of mechanical forces to separate the cellulose fibers from lignin bonds. The most common mechanical pulping technique involves pressing the logs against wet grindstones or passing wood chips between counter-revolving grooved metal discs (refiners) to separate the cellulose fibres. The lignin and other residues are not properly removed in this process. The pulp obtained is of somewhat low quality, which limits the use of this process to less expensive grades of paper such as newsprint. Common mechanical pulping processes are thermo-mechanical pulping (TMP), chemi-thermo mechanical process (CTMP), and semi-chemical pulping consisting of a chemical pre-treatment stage followed by mechanical refining steps (Jukka A Rintala et al. 1994).

Chemical Pulping

In chemical pulping, the cellulose fibers are released from lignin bonds through the application of chemicals. The two popular methods are sulphate(kraft) pulping and sulphite pulping.

Kraft Pulping

It is the most used chemical pulping method, which nearly account for 92% of chemical pulp produced in the world. In kraft pulping, wood chips are cooked with white liquor, which is a mixture of sodium sulfide (Na_2S) and sodium hydroxide ($NaOH$) in a high-pressure digester at elevated temperature.

Kraft pulping produces pulp of long fibers, high strength and very low lignin content. Most mills adopt the kraft method and majority of these kraft mills operate a closed loop system whereby 95-98% of chemicals used in process are recovered and reused (Jukka A Rintala et al. 1994).

Sulphite Pulping

This method use similar equipment to Kraft technique but here the raw materials are cooked in liquor of reactive metal base (usually calcium, magnesium or sodium or ammonium) using H_2SO_4 or Hydrogen sulphite. The cooking liquor solubilizes the wood lignin as lignosulphonic acids in neutral and acidic conditions and produce spent sulphite liquor (Jukka A Rintala, et al. 1994).

The waste stream from the digesters containing lignin, polysaccharides and other sugars, resins, acids etc, is termed as black liquor. Most of the large and some medium mills in India segregate the black liquor for chemical recovery. Mills without chemical recovery systems mix the black liquor with the rest of the wastewater to generate a combined effluent. The major disadvantage of this method is the generation of a dark effluent and also highly dark colored pulp, which requires strong bleaching.

Large Indian mills, whose raw material usually comes from forests, use the Kraft process. Agro-based mills prefer the soda-AQ process while mills making newsprint use mechanical, chemical and chemi-thermomechanical (CTMP) processes (Mohanty 1997). The pulping is generally followed by washing, dewatering and screening processes.

Bleaching Process

Bleaching is a process of treatment of pulp with chemical reagents to increase the brightness of pulp. Bleaching whitens pulps are used for manufacturing writing, printing, and decorative paper.

Mechanical Pulp Bleaching

This process involves altering or masking the lignin present in mechanical pulp, instead of removing it. Bleaching is performed using chemicals that may be oxidizing (H_2O_2), reducing (dithionites) or a combination. The color reduction occurs by altering chromophoric groups arising in lignin of pulp.

Chemical Pulp Bleaching

This process involves removal of lignin from chemical pulp by the application of chemicals. The bleaching improves the brightness of pulp and it is generally performed in multiple stages namely Chlorination stage (C), Extraction stage (E),

Hypo chlorite stage (H), Chlorine dioxide stage (D), Peroxide bleaching stage (P), Oxygen stage (O) and Ozone bleaching (Z). The bleaching chemicals employed in these bleaching stages are respectively chlorine, sodium hydroxide, sodium hypochlorite, chlorine dioxide, hydrogen peroxide, oxygen and ozone (Kumar Swamy 2008).

The chlorination stage, chlorine is used for bleaching chemical pulps and it is mainly responsible for the generation of chlorinated organic compounds. The chlorine converts remaining lignin into chlorinated organic materials such as dioxins and organochlorine compounds. The dioxins and chlorinated organic compounds produced in chlorination stage are reportedly toxic and carcinogenic in nature (P. Singh 2004).

Paper Making Process

In paper making stage, the finished and washed pulp is rolled into sheets of paper in machines. Paper-making consists of stock preparation, forming, pressing and drying in a batch or continuous process. During preparation, the pulp is made more flexible through beating, a mechanical pounding and squeezing process. The pulp is mixed with appropriate fillers such as clay, titanium dioxide, calcium carbonate; sizing agents such as starch, resins and also the pigments at this stage. Forming involves spreading the pulp on a screen. The water is removed by pressing and the paper is left to dry. The paper making process produces a considerable amount of effluent that contains the remains of dyes and washing chemicals used in this process.

WASTEWATER GENERATION AND CHARECTERISTICS

Water Usage Pattern

The paper production process uses a large amount of fresh water, which eventually generates a considerable amount of effluent. It is estimated that the pulp and paper mills consumes 75–225 m^3 of raw water per tonne of paper manufactured. The water consumption in large mills is 125–200 m^3/tonne of paper while small and medium mills consume 75–210 m^3/tonne. Agro-based units consume 125–225 m^3/tonne of paper; in comparison, wastepaper based mills consume only 75–100 m^3/tonne of paper. The water consumption in various stages of paper production is shown in Figure-3. Most agro-based pulp and paper units depend on natural sources of water such as ground water and surface water, river and canals. The cost of water supply varies widely and is in the range of Rs 0.09-3.0 per m^3 (TERI 2003) depending upon the geographical location of the mill.

Effluent Generation

The composition of generated effluent is extremely complex and varies from mill

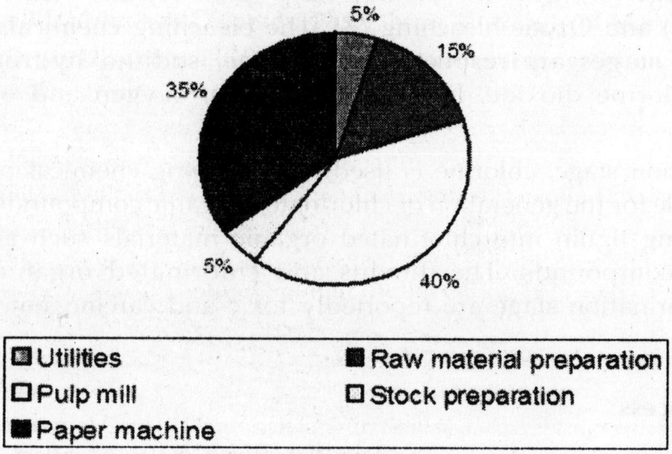

Figure 3: Water consumption in various stages of pulp and paper manufacture (Ansari 2004)

to mill depending on the wood species being pulped, pulping method and bleaching reagents used. The most significant sources of wastewater among various process stages are wood preparation, pulping, pulp washing, screening, washing, bleaching, and paper machine and coating operations (Viraraghavan 2004). The effluent generated in paper mill is characterized as Chipper house effluent, paper mill effluent and recovery plant effluent.

Chipper House Effluent

The wet debarking process generates wastewater containing bark, sand and ligneous materials. The chipper house effluent is low in COD and BOD values. It is found to contain tannins, monomeric phenols, simple carbohydrates and resin compounds (Field et al. 1998).

Paper Mill and Recovery Plant Effluent

The paper mill and recovery plant effluent is originated from pulping, bleaching and paper rolling stages. The composition of effluent is extremely complex and a very little is known about the actual composition of paper mill effluent. The composition generally varies from mill to mill depending on the wood species being pulped, pulping method being adopted, bleaching reagents employed etc. For this reason, the effluent from each mill is characteristically unique in terms of composition (Kumara Swamy 2008).

The pulping process utilizes large amounts of water, which reappear in the form of an effluent. The pulping process generates a high-strength wastewater especially by chemical pulping. This wastewater contains wood debris and soluble wood materials. The pulp fibers obtained from pulping processes consists of various compounds such as lignin, carbohydrate, and extractives, which are hard to biodegrade, and these derivatives are washed away from the fibers during the washing, dewatering, and screening processes (Viraraghavan 2004).

In kraft pulping, about 55% of weight of wood is dissolved in pulping liquor and is found to contain the degradation products of lignin, polysaccharides and wood extractives(Rydholm, 1965).The digester and evaporator condensates produces about 5% of total volume of mill's effluents and are rich in terpenes, ketones, phenols and reduced sulphur compounds. The effluent from sulphite process mainly contains lignosulphonates, carbohydrates, acetic acids, methanol and furfural as major constituents and it also contains small concentrations of formaldehyde, formic acids, acetaldehyde and methyl glyoxal (Jukka A Rintala 1994).

The water consumption in both kraft and sulphite process is high and is about 160-205 tonnes per tonne of pulp. On the other hand, mechanical pulping uses 45-48 tonnes of water per tonne of pulp (Viraraghavan 2004). The wastewater and pollution load generated from individual pulping and papermaking processes is shown in Table 2.

Table 2
Typical wastewater generation and pollution load from pulp and paper industry (Rintala and Puhakka, 1994).

Process	Wastewater (m^3/adt pulp Or paper)	SS (kg/adt pulp)	COD (kg/adt pulp)
Wet debarking	5-25	nr	5-20
Ground wood pulping	10-15	nr	15-32
TMP-unbleached	10-30	10-40	40-60
TMP-bleached	10-30	10-40	50-120
CTMP-unbleached	10-15	20-50	70-120
CTMP-bleached	10-15	20-50	100-180
NSSC	20-80	3-10	30-120
Ca-sulfite (unbleached)	80-100	20-50	nr
Ca-sulfite (bleached)	150-180	20-60	120-180
Mg-sulfite (unbleached)	40-60	10-40	60-120
Kraft-unbleached	40-60	10-20	40-120
Kraft-bleached	60-90	10-40	100-140
Paper making	10-50	nr	nr
Agrobased small paper mill	200-250	50-100	1000-1100

nr—not reported; adt—air dry ton; NSSC—neutral sulphite semi-chemicals.

CHARACTERISTICS OF EFFLUENT

Physicochemical Characteristics

The amount and characteristics of pollutants produced by an individual mill is an important indicator to evaluate the performance of the system. The characteristics of wastewater produced at various pulp and paper making processes varies depending on the process adopted in paper production.

The exact chemical composition of pulp and paper mill effluent is complex and unclear. The chemical compound found in pulp and paper mill effluent are mostly degrading products of lignin, cellulose, hemicellulose and wood extractives (Rydholm. 1965). The lignin degradation products found in the pulp and paper mil effluent include a wide variety of compounds such as manomeric phenols, enol ethers, mercaptides, stilbene, quinone derivatives, chlorinated phenols, acetic acid, formic acid, acetaldehyde, methanol, furfural and methyl glyoxal (Jukka A Rintala et al. 1994). About 300 organochlorine compounds have been identified in effluent while hundred other remain unidentified. The most dangerous of these compounds are chlorophenols such as guaicols, catechols and their transformation products anisoles and verathroles. The most common chlorophenols are extremely toxic and persistant trichlorophenol (TCP) and pentachlorophenol (PCP). Dioxins (Poly chlorodibenzo-p-dioxin or PCDD) and furans (PCDF), which are the most dangerous, chlorinated compounds reported. Apart from dioxins and furans, the other chlorinated compounds found in pulp and paper mill effluent include chloroforms, chloroacetones, aldehydes and acetic acids (Jukka A Rintala et al. 1994).

Absorbable organic halogens (AOX) and total organo chlorine (TOCl) are the other deadly organochlorinated compounds found in effluent. The chlorinated organic compounds are formed during chlorine bleaching stage of paper production. It is estimated that conventional chlorine bleaching of kraft pulp produces about 100-300g of chlorinated phenolic compounds per tonne of pulp (Jokela and Salkinja-Salonen. 1993). In order to reduce the amount of chlorinated organic compounds produced the mills are increasingly adopting use of oxygen bleaching and use of chlorine dioxide bleach in place of chlorine gas. The general characteristics of the paper mill effluent are discussed below and are summarized in Table-3.

Colour and Turbidity

The pulp and paper mill effluent is usually dark colored, foul odoured and turbid in nature. The dark color is caused by the presence of higher lignin and its degradation products. It is estimated that the manufacture of one tonne of bleached paper release about 40-50 kg lignin and its products to waste water (Abassi. 1985).

Most of the colour to the wastewater is contributed by lignin degradation products formed during black liquor stage of pulping. The colour as high as 65,000 units is reported in literature (Singh et al. 1996). The effluent is generally turbid and

Table 3
Characteristics of wastewater at various pulp and paper processes (Viraraghavan, 2004).

Process	Parameters					
	pH	TS (mg/1)	SS (mg/1)	BODs (mg/1)	COD (mg/1)	Colour (Pt–Co)
Large mills (India)	11.0	5250	1233	983	2530	Black
Small mills (India)	12.3	15,120	4890	2628	6145	DB
Digester house	11.69	15,589	23,319	13,088	38,588	16.6[2]
Combined effluent	7.6	3318	2013	103	675	1.0[2]
TMP whitewater	4.7	-	91	1090	2440	-
TMP whitewater	4.7	-	105	1125	2475	-
Kraft mill	8.2	8260	3620	-	4112	4667.5
Pulping	10	1810	256	360	-	-
Kraft mill (unbleached)	8.2	1200	150	175	-	250
Bleaching pulp mill	7.5	-	1133	1566	2572	4033
Bleaching	2.5	2285	216	140	-	-
Pulp and paper	7.8	4200	1400	1050	4870	DB
News air and land paper deinking	8.3	450	400	16	78	-
Paper making	7.8	1844	760	561	953	Black
Paper mill	8.7	2415	935	425	845	DB
Paper machine	4.5	-	503	170	723	243
Paper machine	8.3	-	1032	240	-	-

[2]Unit (Optical Density O.D. at 465 nm); 'DB' means dark brown; 'LY' means light yellow.

opaque due to excessive presence of organic solids. The dark colour and turbidity often has serious consequences on disposal into water bodies.

Total Solids (TSS and TDS)

The total solids in pulp and paper mill effluent consist of both setlable and non-setleble solids. Each category consists of both organic and inorganic substances. Chakravarty et al. revealed that the total solids in pulp and paper mill combined effluent range from 3200-6940mg/l, which is well above the minimum acceptable level (MINAS). The high-suspended solid imparts high COD, BOD and turbidity to effluent.

COD and BOD

The pulp and paper mill wastewater is mostly deficient in dissolved oxygen and associated with extreme values of BOD and COD. The high BOD is caused due to suspended low molecular weight solids. Chakravarty et al (1995) reported that in most cases, the BOD of pulp and paper mill effluent varied between 1200 to 1800 mg/l. This violates of minimum acceptable standard for BOD and also ISI limit for discharge into land for irrigation, which respectively stand at 100mg/l and 500mg/l. The literature shows reports of BOD as high as 16,000mg/l (Singh et al. 1996).

The pulp and paper mill effluents are characterized by high chemical oxygen demand. The lignin and its degradation products not only impart colour and odour to waste water but also increase its BOD and COD values. The pulp and paper mill combined effluents are reportedly contains COD values ranging from 1000 to 4200mg/l (Singh et al 1996). The organic composition and COD characteristics of pulp and paper mill effluents from different processes are shown in Table 4.

Table-4

The organic composition and COD characteristics of pulp and paper mill effluents (Rintala and Puhakka, 1993).

Wastewater	COD (mg COD/l)	Organic Composition (% of COD)	Potential Inhibitory Compounds
Wet Debarking	1300-1400	Tannisns 30-55; monomeric phenols 10-20; simple carbohydrates 30-40; resin compounds 5	Tannins, resin acids
Sulphite Spent Liquor	120000-220000	Lingnosulphates 50-60; carbohydrates 15-25	Not Reported
Sulphite Evaporator Condensate (SEC)	75000-50000	Acetic acid 33-60, mechanol 10-25, fatty acids <10	Sulphur, organic sulphur
Chlorine Bleaching	900-2000	Chlorinate lignin polymers 65-75, methanol 1-27	Chlorinated phenols, resin acids
Kraft Evaporator Condensate (KEC)	1000-33600	Methanol 60-90	Sulphur, resin acids, fatty acids, volatile terpenes
TMP Effluent	1000-5600	Carbohydrates 25-40	Resin acids
CTMP Effluent	2500-13000	Polysaccharides 10-15, lignin 30-40, organic acids 35-40	Resin acids, fatty acids, sulphur

ENVIRONMENTAL CONCERNES AND DISPOSAL NORMS

Effluent Disposal Standards

The Central Pollution Control Board (CPCB) has prescribed the following effluent discharge standards for pulp and paper mills under E (P) Act 1986 (Table 5). In 2003, CPCB constituted a task force on Corporate Responsibility for Environmental Protection (CREP). Under this task force, the following action plan has been formulated and agreed upon by the industries:

Table-5

Effluent discharge standards for pulp and paper mills under E (P) Act 1986

Category	Standard	Note
Large (writing & printing) pulp and paper mills	200 m^3/tonne of paper	Standards of discharge for the large pulp & paper mills established from 1992 onward to meet standard of 100 m^3/tonne of paper
Large (Rayon grade / Newsprint) Pulp and paper mill	150 m^3/T	
Agro-residue based	200 m^3/ tonne of paper	Agro-residue based mills established from January 1992 onward to meet the standards of 150 m^3/tonne of paper
Wastepaper based mills	75 m^3/tonne of paper	Wastewater based mills established from January 1992 onward to meet the standards of 50 m^3/tonne of paper

Large-scale Pulp & Paper

- Wastewater discharge volume <140 m^3/tonne of paper within two years.
- Wastewater discharge volume < 120 m^3/tonnes of paper within four years (for units installed before 1992).
- Wastewater discharge volume <100 m^3/tonne for units installed after 1992.

Small-scale Pulp & Paper

- Wastewater discharge volume < 180 m^3/tonne of paper within three years.
- Recovery of chemicals by installing chemical recovery plants or utilizing black liquor with no discharge from pulp mills within three years or a shift to wastepaper.
- ETPs to be upgraded within one year so as to meet discharge standards
- Utilization of treated effluent for irrigation whenever possible.

The pollution load produced by pulp and paper mills varies from mill to mill. The environmental guidelines on discharge of the pollutants also differ from country to country. The pollutant load discharge guidelines for the pulp and paper industry of some countries are presented in Table 6.

EFFLUENT DISPOSAL AND ITS ENVIRONMENTAL CONSEQUENCES

Pulp mill and bleach plant effluents are highly colored due to polymeric lignin

Table 6

Discharge limits (monthly, semiannual, or annual verges) for bleached kraft pulp (Viraraghavan, 2004).

Country	Parameters				
	SS (kg/adt)	BOD (kg/adt)	COD (kg/adt)	AOX (kg/adt)	References
Canada	9.5-14.5	5.5-30	-	1.4-1.5	TAPPI, 1990
Finland	5-15	6.8-34	90	1.4	TAPPI, 1990
Norway	5	-	90	6	TAPPI, 1990
Sweden	0.3-5.8	7.5-17	39-107	1.5-2	TAPPI, 1990
Belgium	7-14.4	2.3-5.4	22-63	1.5	TAPPI, 1990
France	6.5-10	3.3-30	48-95	-	TAPPI, 1990
USA	3.86 (8.27)	2.41 ()4.52	Reserved	0.272 (0.476)	US EPA, 2000

The US EPA values are monthly average values for new bleached kraft mill. The values in the ( ) are daily maximum allowable.

degradation products and contains chlorinated aromatics. The toxic and environmentally persistent dioxin like compounds resent in the effluent of pulp and paper mill (Bumpus et al. 1988). The pulp and paper industries every year generate more than 7000 billion gallons of highly colored and toxic waste effluents mainly containing high molecular weight, modified and chlorinated lignins (Sundman et al. 1981; Huynh et al. 1985). Most pulp and paper mills dispose the effluent into water bodies such as rivers and canals or onto the barren land.

Aquatic Species

Studies demonstrated a variety of responses in fish populations living downstream of bleached kraft pulp mills These included delayed sexual maturity, smaller gonads, changes in fish reproduction and a depression in secondary sexual characteristics (Munkittrick et al. 1997). The main problem which occurred due to pulp and paper mill was growth of sewage fungus in the river receiving effluents (Webb. 1985).

The dark color and high turbidity due to suspended solids can cause the problems of both water opacity and blanketing of river or lakebeds. Severe blanketing may result in anaerobic decomposition under the blanket releasing hydrogen sulphide into aquatic ecosystems. The dark colour and blanketing can reduce photosynthetic activity in aquatic plants (P. Singh, 2004). This leads to a chain of adverse effects on the aquatic ecosystem as the growth of primary consumers as well as secondary and tertiary consumers are adversely affected (Ruggicro et al. 1989). Therefore, it becomes necessary to remove color and toxicity before they can be accepted into surface waters.

Soil Pollution

Elements such as magnesium, sodium chloride and sulphur, which are also common in pulp mill wastewater can cause nutrient imbalance in crops, increase soil salinity and deteriorate soil structure and ultimately lower crop productivity in long run (Sundari and Kanakarni. 2001).

Crops and Livestock

Pulp and paper mill effluent is also responsible for affecting the quality of crops due to irrigation of polluted water; damages the soil, growth, quality and yield of the crop. Somashekar et al. (1984) in their study on the effect of various industrial effluents observed that paper mill wastewater has inhibiting effect in the germination of crops. Their experiment showed that 18, 14 and 13 percent germination for paddy crop respectively with 25, 50 and 100% waste water. Higher concentration of soluble salts and heavy metals present in pulp and paper mill effluents affect the seed germination of sunflower and maize (Rajamani and Oblisami, 1979; Sahai et al., 1985). The inhibitory effect of the effluent increased not only with the increasing concentration but also with the increasing during of soaking (Kidd and West, 1968).

The paper mill effluents adversely affect the germination of rice seeds (Hari Om et al., 1984). Dutta and Boissya (1997) reported that the germination of rice seeds show a gradual decline at 50% and above concentrations of the effluent, throughout the experimental period, in comparison to control seeds. Mishra and Sahoo (1989) have shown a 44% reduction in shoot weight in 100% paper mill wastewater treated soil.

The farm animals are affected by means of pollutants in the water, soil and air. The pollutant may enter through crops and water consumed by the livestock, as the water borne contaminates accumulate in vegetation and fodder.

WASTEWATER TREATMENT METHODS

Agro residue based pulp and paper mills generally treat combined effluent. The treatment sequence involves equalization, primary settling and clariflocculation followed by secondary biological treatment (anaerobic and/or aerobic). This is followed by the activated sludge process and secondary clarification. Most of the large and a few medium and small paper mills have chemical recovery plants to recover spent pulping chemicals. The treated effluent is disposed of on land, surface water (river) or in drains. The primary sludge is dried in sludge drying beds or lagoons depending upon land availability and is generally sold to board manufacturers. The waste water treatments are broadly categorized into physicochemical and biological methods.

Physicochemical Treatment Methods

Several physicochemical color removal methods such as adsorption, rapid sand filtration, chemical precipitation, membrane processes and electrochemical methods have been developed and reported in literature in the past (Viraraghavan 2004).

The adsorption methods are increasingly being considered for removal of synthetic organic chemicals, color forming organics and disinfection by-products. The different adsorbates commonly used in effluent treatment include activated carbon, processed bone, char powder, activated alumina, magnesia, activated bauxite, fly ash, alum, lime etc (Viraraghavan 2004). Activated carbon is the main adsorbent in full-scale effluent treatment. Other naturally adsorbents are used in special cases. Activated alumina is widely used for removal of fluoride. Silica Gel is used for the separation of hydrocarbons. The polymeric resins and carbonized resins are often employed for improved removal of organic compounds from effluent.

The membrane techniques require pretreatment and requires large capital investment. Membrane fowling is another problem associated with this method. Adsorption and membrane process are efficient but expensive (Manjunath and Mehrotra. 1981). The application of electrochemical method is another way to treat the wastewater from the cellulose paper production (Christoskova and Lazarov. 1988). This method guarantees high treatment efficiency, but its effectiveness depends on the type of electrodes, the construction of electrocoagulators and the condition under which the process is run.

Chemical precipitation using alum, ferric chloride and lime has been studied extensively (Lathia and Joyce, Joyce et al. 1979), Despite the short detention time and low capital cost, there are some drawbacks reported, such as high cost of chemicals for precipitation and pH adjustment, voluminous sludge production due to heavy dosages, dewatering and disposing of generated sludge and high residual cation levels, The chemical precipitation methods are cheap but produce a large quantity of sludge and do not completely remove toxicity.

The chemical aspect of colour removal of effluent from pulp and paper industry is very important. The use of calcium hypochlorite (1-2% as available chlorine) during alkaline extraction reduced the colour of effluent by 84% without affecting the quality of pulp. The use of chlorinated backwater (having 0.8% residual chlorine) during brown stock washing reduced the colour of effluent by 60% without affecting the quality of pulp. The combination of alum, lime and magnesium sulphate in presence of ferric acid chloride reduced the colour, BOD and COD by 97%, 68% and 52%, respectively. The combination of alum, calcium hypochlorite and ferrous sulphate in presence of chlorine water was most effective and reduced the colour, BOD and

COD by 97, 71 and 64% respectively. The treatment options that have been explored till now are not cost-effective at plant level and no completely efficient method is currently available.

Biological Treatment Methods

Biological methods have the potential to eliminate or reduce the problems associated with physicochemical methods. Several studies have been carried out concerning the decolourization and treatment of such wastewaters by biological methods. Color of paper mill effluent is largely due to lignin and lignin derivatives and polymerized tannins, which are resistant to degradation due to the presence of carbon-to-carbon biphenyl linkages. It is reported that lignin and lignin derivatives are biodegradable by some of microorganisms under proper environmental conditions. Numerous bacteria have been reported to decompose lignins and lignin derivatives and some of these being *Pseudomonas spp., Flavobacteria, Xanthomonas spp., Bacillus spp., Aeromonas spp., Cellulomonas spp., Chromobacrtia,* etc (Ebtesam El-Bestawy et al 2008). Although numerous bacteria can decompose monomeric lignin structure models, only a few strains are able to attach lignin derivatives obtained from different pulping processes.

Many of the past studies have focused on screening, identifying, and evaluating the ability and effectiveness of fungi on degrading lignins in situ and in vitro. A variety of fungi has been proved to be lignin degraders and are classified into white-rot, soft-rot, and brown-rot fungi based on the type of wood decay carried out by these organisms (Ebtesam El-Bestawy et al, 2008).

The white rot fungi are a group of basidiomycetes that possess an active lignolytic enzyme system, which are most efficient of the microorganisms that degrade lignin and its modified forms (Kersten et al, 2006). These fungi do not use lignin as a carbon source for their growth but use it as a secondary metabolite which is not required for their growth. The lignin degradation by white rot fungi is extensively studied and degradation is caused by three extracellular phenol oxidases namely lignin peroxidases (LiP), manganese peroxidases (MnP) and laccases (Lac)(Peng Wang et al 2008). In addition to degrading lignin, these fungi are also capable of degrading a variety of environmentally persistent pollutants such as chlorinated aromatic compounds, heterocyclic aromatic hydrocarbons, synthetic high polymers and various dyes (Ohkuma et al, 2001).

Several authors reported on the capacity of different fungal species to remove color from kraft mill effluent (Gokcay and Dilek, 1994; Duran et al., 1994; Sakurai et al., 2001). Prasad and Gupta (1997) reported on a substantial reduction of color and COD by the use of white rot fungi *T. Versicolor* and *P. Chrysosporium*. Saxena and Gupta (1998) showed that white-rot fungi *P. Chrysosporium* in combination with other white-rot fungi (*P. sanguineus, P. ostreatus and H. annosum*) and with the use of the surfactants were able to remove color, COD, and lignin content. Choudhury et al. (1998) found that lignin, BOD, COD and color removal were achieved to the extent of 77%, 76.8%, 60%, and 80%, respectively, by the fungal specie *Pleurotus ostreatus*. Taseli and Gokcay (1999) isolated fungal specie (*Pencillium sp.*) which was able to remove 50% of the AOX, and color from the soft-wood bleachery effluents in a contact time of 2 days. Zhang et al. (2000b) showed that fungus such as *T. versicolor* and fungal culture filtrate (FCF) obtained from these organisms were able to efficiently degrade the dissolved and colloidal substances. The other white rot fungi reported to degrade effluent color under optimum conditions include *Tinctoporia borbonica, Schizophyllum commune, Aspergillus fumigatus, Pleurotus ostreatus* among others (P. Singh, 2004).

Aerobic Treatment

Biological oxidation is the most widely used techniques to remove BOD and chlorinated organics because of its effectiveness and low cost.

Activated Sludge Process

Shere and Daly (1982) claimed that TMP wastewater was readily degradable by the activated sludge process. Gergov et al. (1988) reported that biological treatment by the activated sludge process removed 79-95% of chlorophenolics.

Junna and Ruonala (1991) reported 90% BOD_7, 70% COD, 40–60% AOX, and 60–95% chlorinated phenols removal by the activated sludge process. Knudsen et al. (1994) reported a high reduction of BOD and soluble COD by a two-stage activated sludge process. Chandra (2001) reported efficient removal of color, BOD, COD, phenolics, and sulfide by microorganisms such as *Pseudomonas putida, Citrobacter sp., and Enterobacter sp.* in the activated sludge process. High removals of BOD, COD, AOX, and chlorinated phenolics have been achieved in the activated sludge process (Saunamaki, 1997; Schnell et al., 2000a). Kennedy et al. (2000) reported that the activated sludge was successful in removing nearly all detectable Microtoxk toxicity from bleached kraft pulp mills at low level whereas the PACTk was slightly better in removing highly toxic concentrated effluents.

Aerated Lagoons

Leach et al. (1977) reported the biodegradation of several compound representing the major categories of toxicants in laboratory batch aerated lagoon. Eriksson and Kolar (1985) have shown that the high molecular mass fraction in bleach effluent cannot be degraded in an aerated lagoon. Stuthridge et al. (1991) reported 65% removal of AOX from bleached kraft pulp and paper mill effluent. Junna and Ruonala (1991) reported removal of BOD_7 ranging between 50% and 75% and chlorinated phenolics 10–50% by an aerated lagoon. Chernysh et al. (1992) reported large variations in AOX and TOC removal in a controlled batch study of bleached kraft effluent in an operating lagoon under both aerobic and anaerobic conditions. Welander et al. (1997) reported COD removal of 30–40% in a full scale lagoon and

60–70% in a pilot-scale plant. Achoka (2002) reported that an oxidation pond removed chemical compounds greater than 50%. Schnell et al. (2000a) reported removals of BOD, AOX, chlorinated phenolics, and polychlorinated phenolics respectively from an aerated lagoon.

Aerobic Biological Reactors

Franta (1994) and many other authors have reported high removals of organic pollutants of kraft mill wastewater by sequencing batch reactor (SBR) treatment. Removal of COD by a moving bed biofilm reactor (MBBR) had been demonstrated (Jahren et al., 2002; Borch-Due et al., 1997). Magnus et al. (2000c) reported 93% and 65% removal of BOD and COD, respectively by a biological compact reactor. Asselin et al. (2000) concluded that suspended carrier biofilm reactor (SCBR) was highly efficient in removing chronic toxicity from the effluent. Rovel et al. (1994) achieved 76%, 62%, 81%, and 48% removal of BOD, COD, SS, and AOX, respectively, using a biofilter.

Anaerobic Treatment

Anaerobic digestion is a process frequently employed for the secondary treatment of industrial wastewater and is considered more suitable to treat high strength organic effluents. The commonly used anaerobic processes to treat pulp and paper mill effluents are Anaerobic filter, Upflow sludge blanket (UASB), Fluidized bed, Anaerobic lagoon, and Anaerobic contact reactor methods.

Poggi-Varaldo et al (1996) investigated the pretreatment of the kraft mill black liquor and they reported that continuous anaerobic treatment feasible at low to medium loading rates, with a total COD removal of 48–80% and biodegradable COD reduction of 87–96%. Jahren et al. (1999) compared anaerobic and aerobic treatment for TMP mill effluent and found that 84% and 86% removal of COD from anaerobic and aerobic treatment systems, respectively, was achieved. Rajeshwari et al. (2000) reported that chlorine bleaching effluents were not suitable for anaerobic treatment due to their low biodegradability and presence of toxic substances that affects methanogens. Jackson-Moss et al. (1992) found 50% removal of COD and color by anaerobic biological granular activated carbon. Dufresne et al. (2001) observed that undiluted foul condensates at Windsor Mill were toxic to anaerobic biomass. Peerbhoi (2000) investigated anaerobic treatability of black liquor by a UASB reactor and author concluded that anaerobic biological treatment of black liquor was not feasible, as the pollutants were not readily degradable. Rajeswori et al. (2000) reported a 50% reduction of BOD of debarking wastewater by a fluidized bed reactor. Thompson et al. (2001) reported that COD removal efficiency of 80% was constantly achievable but the residual COD was around 800 mg/ l meaning that additional treatment was essential. Schnell et al. (1992) concluded that anaerobic treatment systems were less suitable for treatment of sulfite-spent liquor compared to an aerobic system.

Anaerobic biological treatment also effectively destroys chlorophenolic compounds, mutagenicity and acute toxicity in the bleaching effluent (Bajpai et al., 1997). It has many potential advantages in comparison to aerobic treatment such as lower sludge production, lower chemical consumption, smaller land requirement due to smaller reactors and energy production in the form of methane (Minqing Yang (2008). Minqing Yang (2008) reported the dependence of anaerobic degradability of effluent on the type of wood used in pulp production. The hardwood was shown to have better treatment than softwood effluent from the same process.

Integrated Treatment Processes

These methods combine one or more of the treatment techniques either in combination or in a sequence. Kabdash et al. (1996) showed that a combination of chemical and biological methods (bioferic) resulted in 40–50% additional removal of COD compared to the activated sludge system. Jahren and Oedegaard (1999) found that Kaldnes (anaerobic followed by aerobic) moving bed biofilm reactor removed about 60% of soluble COD from TMP whitewater. A combination of ozone and biofilm reactor removed 80% COD (Helble et al., 1999). Zhang et al. (2000a) examined the removal of most of the detrimental organics from whitewater by combined enzyme and fungal treatment. Duncan and Thia (1992) suggested a combined anaerobic–aerobic treatment system was to treat bleached kraft pulp and paper mill effluents. Lescot and Jappinen (1994) showed that a combination of an aerated lagoon and a secondary clarifier was able to remove 87%, 96%, 65%, 53%, and 22% of SS, BOD_7, COD, AOX, and color, respectively from a bleached kraft mill effluent. Carlson et al. (2000) reported that 77% and 98–99% removal of COD and BOD respectively by a system comprising of an anoxic selector, an aerated basin, and a secondary clarifier in series. A combination of physicochemical, biological, and effluent polishing in the aerated lagoon removed 98–99% BOD, 91% COD, 97% SS, and 90% color of a pulp and paper mill in Brazil (Foelkel, 1989). Rusten et al. (1994) reported that a combination of a biofilm reactor followed by one anaerobic and two aerobic reactors was found to remove 50% COD, 80–90% BOD7, 50% AOX. Shaw et al. (2002) showed that a combination of aerobic reactor followed by anaerobic reactor removed 94% color, and 66% TOC. W. De los Santos Ramos et al (2009) showed that the remediation of lignin and its derivatives from pulp and paper industry wastewater by the combination of chemical precipitation and ozonation resulted in efficient color and COD reduction.

Discussions

The literature review shows that an internal process change is one of the options to be adopted by the pulp and paper industry to reduce the pollution at the source. The biotechnological applications such as genetic modification of plant, biopulping, and biobleaching to reduce chlorinated organic compounds as an emerging technology for internal pollution control. Enzyme treatment for pulp dissolving, improving tensile properties by treating mechanical pulp with white rot organisms and enzymatic beating of chemical pulps, hemicellulose, and decolorization by white rot fungi were given as possible biotechnological options.

Among the various treatment processes currently used for pulp and paper effluent treatment, only a few are commonly adopted by pulp and paper industry especially for tertiary treatment. Some of the treatment processes such as ozonation, fenton's reagent, adsorption, and membrane technology are efficient but are more expensive. Sedimentation is the most commonly adopted process by the pulp and paper industry to reduce suspended solids. Flotation is also commonly used in the pulp and paper industry but most of the time as a tertiary treatment. Coagulants are a preferred option for removing turbidity and color from the wastewater and they are also capable in reducing COD, TOC, and AOX to some extent. Among the coagulants, modified chitosan showed the highest performance for color and TOC removal. Polyelectrolytes are better than alum and they produce less sludge and pose lesser problems with sludge dewaterability than alum. Adsorption processes are useful to remove color, COD, and AOX. They are rather expensive and it is not known whether the pulp and paper industry are employing them widely. However, laboratory-scale experiments are usually reported. Activated charcoal, fuller's earth, and coal ash showed better results for color removal. Activated coke alone was able to remove 90% of the COD, AOX, DOC, and color.

Chemical oxidants such as (ozone + photocatalysis), and (ozone + UV) are reported to be efficient in removing COD and TOC and color. However, it is rather expensive (Perez et al., 2002b). Ozonation is not commonly adopted in most countries. Membrane filtration processes are efficient in reducing over 90% of color, TSS, and AOX in most of the cases but the fouling of membranes is a problem in the case of soft wood effluent.

In secondary treatment processes, activated sludge is the most commonly used. UASB and fluidized beds are also gaining in use recently. The problem with activated sludge is sludge bulking. Reported results have shown that activated sludge can remove all types of the pollutants pertaining to the pulp and paper industry. However, the removal of AOX is below 50%, BOD around 95% in most of the mills, and COD removal averages around 70%. This system is also efficient in removing chlorinated phenolic compounds (>75%) most of the time. Dalentoft and Thulin (1997) reported that Kaldnes (anaerobic + aerobic) process in series with an activated sludge, could be an efficient, stable, and a competitive combination process, considering both investment and operating costs. Aerated lagoons are efficient in removing BOD over 95% in most of the reported results. COD removals are moderate between 60% and 70%, AOX around 50%, and a high removal (85%) for chlorinated phenolics. Anaerobic contact reactors are efficient in removing biodegradable organic compounds such as BOD, and COD. The performance data from various mills showed that anaerobic contact reactors were able to remove over 90% of BOD and 65% of COD in most of the cases. Anaerobic filters and fluidized bed reactors are suitable in reducing organic pollutants only. Both the reactors achieve almost same efficiency in terms of BOD (>80%), and COD (>70%) removal. UASBs are able to remove over 80% of BOD and 50–80% of COD in most of the mills. Fungi are efficient in removing especially color and COD from the pulp mill wastewater. Removal of color using white rot fungi was above 80% in most of the reported cases and COD removal was above 75%. The reported results have shown that high removals are achieved in the case of the combination of two or more physicochemical processes or combination of physicochemical and biological processes. The confirmation of the reported results, their applicability in the real field, and economic evaluations are very important in adopting the process. For example, the anaerobic treatment process for pulp and paper mill effluents is still in an initial application phase. Bajpai (2000) showed that activated sludge was almost twice as expensive as anaerobic reactors.

Conclusions

Based on above studies the following conclusions are drawn:

1. The pulp and paper mills use various plant bio-resources as raw materials such as bamboo, eucalyptus, agricultural waste residues such as rice straw, wheat, sarkanda grass, bagasse, jute/rags, etc as the raw materials. However, the raw materials used vary from mill to mill depending on the end product desired.
2. The chemical compounds found in pulp and paper mill effluent are mostly degrading products of lignin, cellulose, hemicellulose and wood extractives such as manomeric phenols, enol ethers, mercaptides, stilbene, quinone derivatives, chlorinated phenols, acetic acid, formic acid, acetaldehyde, methanol, furfural and methyl glyoxal. About 300 organochlorine compounds have been identified in effluent while hundred other remain unidentified.
3. Color is imparted to effluent by chromophoric compounds such as lignin and tannin, and the decolorization is efficiently achieved through fungal treatment, coagulation, chemical oxidation, and ozonation techniques.
4. Chlorinated phenolic compounds are produced during pulp bleaching stages. Chlorinated phenolic compounds and AOX can be removed by adsorption, ozonation and membrane filtration techniques. However, more studies are needed on the removal of AOX and chlorinated phenolic compounds.
5. Physical and chemical processes are quite expensive to remove only high molecular weight chlorinated lignins, colour, toxicants, suspended solids and COD. However, BOD and low molecular weight compounds like alcohols and acids appreciably are not removed efficiently by these methods.
6. Aerobic processes such as Activated sludge process and Areated logoons are effective in removal of BOD and chlorinated phenolic compounds. The anaerobic treatment of high strength wastewater is not effective as it leaves high residual COD.

7. Both aerobic and anaerobic treatment systems are feasible to treat wastewater from all types of pulp and paper mills except that bleaching kraft effluents are less suitable for treatment by anaerobic means, as they are more toxic to anaerobic bacteria.
8. A combination using an anaerobic process followed by an aerobic treatment system is a better option, as it can make use of the advantages of both the treatment processes. Combinations of physicochemical and biological treatment processes with optimization of the process provide a long-term solution for pulp and paper mill effluent treatment and are the need of the hour.

References

Abassi, A. (1985). Occurrence, toxicity and treatment of lignin in pulp and paper Effluents: State of Art, *J. Env.*, 65, 1-8.

Achoka J.D. (2002). The efficiency of oxidation ponds at the kraft pulp and paper mill at Webuye in Kenya, *Water Res.*; 36, 1203-12.

Asselin C., Collin D., Graff S. (2000). Effluent treatment for chronic toxicity removal with the suspended carrier biofilm reactor. *Tappi International Environmental Conference and Exhibit. Denver Co.*, Vol. 2. Norcross, GA 30092, USA: Technical Association for Pulp and Paper Industry (TAPPI), p. 805–11.

Bumpus, J.A., Aust. S.D. (1988). Biodegradation of Environmental pollutants by the White rot fungus Phanerochaete Chrysosporium involvement of the lignin degrading system. *Bio Assay*, 6, 1143-1150.

Bajpai, P. and Bajpai, P.K. (1997). Reduction of organochlorine compounds in bleach plant effluents; K.EL(Ed.), *Advances in Biochemical Engineering and Biotechnology*, 57, 213-259.

Bajpai P. (2000). Treatment of pulp and paper mill effluents with anaerobic technology. Randalls Road, Leatherhead, UK, *Pira International.*

Borch-Due A., Anderson R., Opheim B. (1997). Treatment of integrated newsprint mill wastewater in moving bed biofilm reactors, *Water Sci. Technol.*, 35(2–3):173– 80.

Chakravarty, K.R., Singanam M., and Somashekara, K.R. (1995). A correlation study of physico-chemical characteristics of paper mill effluents, *Ind. J. Env. Poll.*, 16(1), 46-49.

Christoskova, S.G., and Lazarov, L.D. (1988). Electrochemical method for purification and dissolution of cellulose paper industry wastewater, *Environ. Prot. Eng*, 14(3-4), 69-76.

Carlson B.L., Ericsson T., Loyblad R., Persson S., Simon O. (2000). The reconstruction of an aerated lagoon to a long-term aerated sludge (LAS) plant at Sodra cell, Monsteras kraft pulp mill. *Tappi International Environmental Conference and Exhibit, Denver, CO*, Vol- 1. Norcross, GA 30092, USA: Technical Association for Pulp and Paper Industry (TAPPI), p.363– 71.

Chandra R. (2001). Microbial decolourisation of pulp mill effluent in presence of nitrogen and phosphorous by activated sludge process, *J Environ Biol.*, 22(1), 23 –7.

Chernysh A., Liss N.S., Allen G.D. (1992). A batch study of the aerobic and anaerobic removal of chlorinated organic compounds in an aerated lagoon, *Water Pollut Res. J. Can.*, 27(3), 621– 38.

Choudhury S., Sahoo N., Manthan M., Rohela R.S. (1998). Fungal treatment of pulp and paper mill effluents for pollution control, *J. Ind. Pollut Control*, 14(1), 1– 13.

Dalentoft E., Thulin P. (1997). The use of the kaldnes suspended carrier process in treatment of wastewaters from the forest industry, *Water Sci. Technol.*, 35(2 – 3), 123–30.

W. De los Santos Ramos, T. Poznyak, I. Chairez, I. Córdova R. (2009). Remediation of lignin and its derivatives from pulp and paper industry wastewater by the combination of chemical precipitation and ozonation, *Journal of Hazardous Materials*, 169, 428–434.

Dufresne R., Liard A., Blum S.M. (2001). Anaerobic treatment of condensates: at a kraft pulp and paper mill, *Water Environ. Res.*, 73(1), 103–9.

Duncan A., Thia B. (1992). Treatment of pulp and paper mill effluent, *Aust. Biotechnol.*, 2(4), 235.

Duran N., Esposito E., Innicentini-Mei L.H., Canhos P.V. (1994). A new alternative process for kraft E1 effluent treatment. *Biodegradation;* 5: 13–19.

Dutta, S.A. and Boissya C.L. (1997). Effect of Paper mill effluent on germination of rice seed (Oryza sativa L. va Masuri) and growth behavior of its seedlings; *J. of Indian Pollution control* 13(1); 41-47.

Ebtesam El-Bestawy, Ibrahim El-Sokkary, Hany Hussein and Alaa Farouk Abu Keela (2008). Pollution control in pulp and paper industrial effluents using integrated chemical–biological treatment sequences; *Journal of Industrial Microbiology and Biotechnology*, doi: 10.1007/s10295-008-0453-3.

Eriksson, K. E. and Kolar, K. C. (1985). Microbial degradation of chlorolignins, *Environ. Sci. Technol.*, 19; 1086-1089.

Field, J. A, De Jong. E., Feijooo-Casta, G, de Bont. J.A.M. (1993). Screening for lignolytic fungi applicable of the biodegradation of xenobiotics, *Trens Biotechnol*, 11; 44-49.

Foelkel C. (1989). Advanced waste treatment, *Pulp Pap*, 63(4), 173–4.

Gergov, M., Priha, M., Talka. E., Valltila, O, Kangas. A and Kukkonen. K. (1988). Environmental conference, *Proc. Tappi Press,* Atlanta, p443.

Gokcay F.C., Dilek F.B. (1994). Treatment of effluents from hemp-based pulp and paper industry (2) biological treatability of pulping effluents. *Water Sci. Technol.*, 29(9), 165–8.

Hari Om, Nepal Singh and Mugank, S. Arya (1984). Combined effect ofwastes of distillaryand suger mill on seed germination; Seedling growth and Biomass of okra (Abelmoschus esculeentus(L)(Moench), *J. Environ. Biology*, 15(3), 171-175.

Helble A., Schalyer W., Liechti P.A., Jenny R., Christian M.H. (1999). Advanced effluent treatment in the pulp and paper industry with a combined process of ozonation and fixed bed biofilm reactors, *Water Sci. Technol.*, 40(11 –12), 343–50.

Huynh, V.B., Chang, H.M., Joyce, T.W., and Kirk, T.K. (1985). Dechorination of chloroorganics by a white rot fungus, Tappi. J, 68, 98-102.

Jackson-Moss C.A., Maree J.P., Wotton S.C. (1992). Treatment of bleach plant effluent with the biological granular activated carbon process, *Water Sci. Technol.*, 26(1–2), 427–34.

Jahren S.J., Oedegaard H. (1999). Treatment of thermomechanical pulping (TMP) whitewater in termophilic anaerobic– aerobic moving bed biofilm reactors, *Water Sci. Technol.*, 40(8), 81–90.

Jahren S.J., Rintala J.A., Odegaard H. (1999). Evaluation of internal thermophilic biotreatment as a strategy in TMP mill closur, *Tappi J.*, 82(8), 141–9.

Jahren J.S., Rintala J.A., Odegaard H. (2002). Aerobic moving bed biofilm reactor treating thermomechanical pulping whitewater under thermophilic conditions, *Water Res. 36,* 1067–75.

Jokela, J.K., Laine, M., Ed. M., and Salkinja-Salonen, M. (1993). *Environ. Sci. Technol.*, 27, 547.

Rintala, J.A., and Puhakka (1994). Anaerobic treatment in pulp and paper mill waste mamagement: A review, *Bioresource Technol.*, 47, 1-18.

Junna J., Ruonala S. (1991). Trends and guidelines in water pollution control in the Finnish pulp and paper industry, *Tappi J.*, 74(7), 105–11.

Kabdash I., Tunay O., Eldem N. (1996). Treatability of wastepaper pulping process wastewater. *Proceedings of the Industrial Waste Conference, West Lefayette, USA*, Vol. 51. Chelsea, Michigan 68118, USA, *Ann Arbor Press Inc*, p. 645– 50.

Kennedy K.J., Graham B., Droste R.L., Fernandes L., Narbaitz R. (2000). Microtox and Ceriodaphnia dubia toxicity of BKME with powdered activated carbon treatment, *Water SA*, 26(2), 205–16.

Kersten P., Cullen D. (2006). Extracellular oxidative systems of the lignin-degrading basidiomycete *Phanerochaete chrysosporium*, *Fungal Genet Biol*, 44, 77–87.

Kidd F. and West C. (1968). Physiological predetermination. The influence of physiological condition of seed upon the course of growth and yield, *Ann. Appl. Biol*, 5, 1-10.

Knudsen L., Pedersen J.A., Munck J. (1994). Advanced treatment of paper mill effluents by a two-stage activated sludge process, *Water Sci. Technol.* 30(3), 173 –81.

Kumara Swamy, N. (2008). Enzymatic Degradation of Pulp and Paper Mill Effluent, *Master's Thesis*, Vinayaka Mission University, Salem Tamilnad, India.

Lathia, S.G., Joyce, T.W. (1978). Removal of color from carbonate pulping effluent; The calcium magnesium coagulation process; *TAPPI J.*, 61(10), 67-70.

Leach, J.M. Muellar, J.C. and Walden, C.G. (1997). CPAR Report No. 408-2, Canada Forestry Service, Ottawa, Canada.

Lescot J.C., Jappinen H. (1994). Effluent treatment in pulp and paper mills: present technology and development trends, *Appita J.*, 47(4), 330– 2.

Magnus E., Hoel H., Carlberg G.E. (2000c). Treatment of an NSSC effluent in a biological high-efficiency compact reactor, *Tappi J.*, 83(1), 149– 56.

Manjunath, D.L., and Mehrotra, I. (1981). Removal of reactive dyes using alum, lignin sludge; *Indian J. Environ Health*, 23(4), 309-315.

Minqing (Ivy) Yang (2008). Anaerobic treatment of In-Mill wastewaters collected from a pulp and paper mill, *Master's Thesis*, University of Toranto;

Mishra P.C. and Sahoo, G. (1989). Agropotentiality of paper mill wastewater.in: *Soil pollution and soil organisms* (Ed P.C Mishra), Ashish Publishing House, New delhi, pp. 97-120.

Munkittrick, K.R., Serveos, M.R., Carey, J.H. and Van der Kraak, G.J. (1997). Environmental Impacts of pulp and paper wastewater, *Water Sci. Technol.*, 35, 329-338.

Ohkuma, M., Maeda, Y., Johima and Kudo, T. (2001). Lignin degradation and role of whiterot fungi: Study on an efficient symbiotic system in fungus-growing termites and its application to bioremediation, *RIKEN Dev.*, 42, 39-42.

Peerbhoi Z. (2000). Treatability studies of black liquor by UASBR-Ph.D. Thesis. University of Roorkee, India.

Perez M., Torrades F., Domenech X., Peral J. (2002a). Removal of organic contaminants in pulp effluents by AOPs: an economic study, *J. Chem. Technol. Biotechnol.*, 77, 525– 32.

Peng Wang, Xiaoke Hu, Sean Cook, Maria Begonia, Ken S. Lee and Huey-Min Hwang (2008).

Effect of culture conditions on the production of ligninolytic enzymes by white rot fungi *Phanerochaete chrysosporium* (ATCC 20696) and separation of its lignin peroxidase, *World Journal of Microbiology and Biotechnology, 24(10), 2205-2212.*

Poggi-Varaldo H.M., Estrada-Vazquez C., Fernandez-Villagomez G., Esparza-Garcia F. (1996). Pretreatment of black liquor spills effluent. *Proceedings of the Industrial Waste Conference, West Lafayette, USA*, 51, 651–61

D.Pokhrel and T. Viraraghavan (2004). "Treatment of pulp and paper mill wastewater-A review", *Science of the Total Environment*, 33, 37-58.

Prasad G.K., Gupta R.K. (1997). Decolourization of pulp and paper mill effluent by two White-rot fungi, *Indian J. Environ. Health*, 39(2), 89–96.

Rajamani G. and Oblisami G. (1979). Effect of paper factory effluent on soil and crop plant, *Ind. J. Env. Health*, 21, 121-130.

Rajeshwari K.V., Balakrishnan M., Kansal A., Lata K., Kishore V.V.N. (2000). State of the art of anaerobic digestion technology for industrial wastewater treatment, *Renew Sustain Energy Rev.* 4(2), 135–56.

Rovel J.M., Trudel J.P., Lavalle P., Schroeter I. (1994). Paper mill effluent treatment using biofliltration, *Water Sci. Technol.*, 29(10–11), 217– 22.

Ruggicro, P., Sarkar, J.M., and Bollag J.M. (1989). *Soil Sci.*, 147, 361.

Rusten B., Mattsson E., Due B.A., Westren T. (1994). Treatment of pulp and paper industry wastewaters in novel moving bed biofilm reactors, *Water Sci. Technol.*, 30(3), 161 –71.

Rydholm. S.A. (1965). Pulping process, *Inter Science Publisher, Newyork.*

Sahai R.N., Shukla N., Jabeen S. and saxena P.K. (1985). Pollution effect of distillery waste on the growth behavior of phaseolus radiatus, *Environ. Pollution*, 37, 245-253.

Sakurai A., Yamomoto T., Makabe A., Kinoshita S., Sakakibara M. (2001). Removal of lignin in a liquid system by an isolated fungus, *J. Chem. Technol. Biotechnol.*, 77, 9–14.

Saunamaki R. (1997). Activated sludge plants in Finland, *Water Sci. Technol.*, 35(2–3), 235–43.

Saxena N., Gupta R.K. (1998). Decolourization and delignification of pulp and paper mill effluent by white rot fungi, *Indian J. Exp. Biol.*, 36, 1049–51.

Schnell A., Hall E.R., Skog S. (1992). Anaerobic and aerobic treatability of high-yield sulphate spent liquor, *Water Pollut Res., J. Can.*, 27(3), 601– 20.

Schnell A., Steel P., Melcer H., Hodson P.V., Carey J.H. (2000a). Enhanced biological treatment of bleached kraft mill effluents: I. Removal of chlorinated organic compounds and toxicity. *Water Res.*, 34(2), 493– 500.

Shaw C.B., Carliell C.M., Wheatley A.D. (2002). Anaerobic/aerobic treatment of coloured textile effluents using sequencing batch reactors, *Water Res.*, 36, 1993– 2001.

Shere S.M. (1982). Daly P.G. High rate biological treatment of TMP effluent, *Pulp Pap. Can.*, 83n, 61– 6.

Singh, P. (2004). Ph.D thesis, G.B. Pant University of Agriculture and Technology, Pantnagar, India.

Singh. R.S., Marwaha, S.S., and Khanna, P.K. (1996). Characteristics of pulp and Paper mill effluents, *J. Ind. Poll Cont.*, 12(2), 163-172.

Somshekar R.K., Gowda M.T.G., Shettigar S.C.N. and srinath K.P. (1984). Effect of industrial effluents on crop plants, *Indian J. Environ. Health*, 26(2), 136-146.

Sundman. G., Kirk. T.K., and Chang, H.M. (1981). Fungal decolorization of kraft bleached effluent. Fate of chromophoric material; *TAPPI J.*, 64; 144-148.

Sundari. S., and Kanakarnim, M.S.P. (2001). The effect of pulp unit effluent on agriculture, *J. of Industrial Poll Control*, 17(1), 83-97.

Taseli B., Gokcay C.F. (1999). Biological treatment of paper pulping effluents by using a fungal reactor, *Water Sci. Technol.*, 40(11– 12), 93–9.

Thompson G., Swain J., Kay M., Forster C.F. (2001). The treatment of pulp and paper mill effluent: a review. Bioresour Technol 77(3), 275– 86.

Webb L. (1985). An investigation into the occurrence of sewage fungus in rivers containing paper mill effluents. 1: Review of previous research, *Water Res.*, 19, 947-959.

Welander T., Ericsson T., Gunnarsson L., Storlie A. (2000). Reduction sludge production in biological effluent treatment by applying the LSP process. *Tappi International Environmental Conference and Exhibit*, Vol. 2, p. 757–63.

Zhang X., Stebbing D., Soong J., Saddler J.N. (2000a). The Removal of detrimental dissolved and colloidal substances by a combined fungal and enzyme treatment system, *86th Annual Meeting, Montreal, Quebec, Canada;* Pulp and Paper Technical Association of Canada; p. B99–B102.

Zhang X., Stebbing D.W., Saddler J.N. (2000b). Enzyme treatment of the dissolved and colloidal substances present in mill white water and the effects of the resulting paper properties,. *J. Wood Chem. Technol.*, 20(3):321–35.

4

Antibacterial Activities of Plant Extracts

I. Bhattacharjee, S.K. Chatterjee and G. Chandra

Abstract

India is blessed with two mega centres of biodiversity: the Hindustan Centre of Origin and the Central Asia Centre of Origin. This biodiversity is mainly distributed in Western Ghat, North Eastern India and the Himalayan Region. The number of higher plant species (angiosperms and gymnosperms) on this planet is estimated to be 250,000, with a lower level at 215,000 and at an upper level of 500,000. Of these, only about 6% have been screened for biologic activity, and a reported 15% have been evaluated phytochemically. Modern innovative techniques have facilitated the process of extraction, isolation and identification of plant materials. This chapter deals with phytocompounds, used in traditional medicine, that are a basis of both regional primary health care and raw material for pharmaceutical products with special reference to antibacterials.

Introduction

Traditional medicine has a long history of serving peoples all over the world (Schultes and Raffauaf, 1990; Arevalo, 1994; Estrella, 1995; Lossio et al., 1995; Desmarchelier and Schaus, 2000; Clavo et al., 2003; Reynel et al., 2003; Tshikalange et al., 2005). Fossil records date human use of plants as medicines at least to the middle Paleolithic age, some 60,000 years ago (Solecki, 1975). From that point the development of traditional medical systems incorporating plants as a means of therapy can be traced back only as far as recorded documents of their likeness. The goals of using plants as sources of therapeutic agents are (a) to isolate bioactive compounds for direct use as drugs, eg., digoxin, digitoxin, morphine, reserpine, taxol, vinblastine, vincristine; (b) to produce bioactive compounds of novel or known structures as lead compounds for semisynthesis to produce patentable entities of higher activity and/ or lower toxicity e.g., metaformin, nabilone, oxycodon (and other narcotic analgesics) taxotere, teniposide, verapamil and amiodarone, which are based, respectively, on galegine, morphine, taxol, podophyllotoxin, khellin; and (c) to use agents as pharmacological tools, e.g., lysergic acid diethylamide, mescaline, yohimbine; and d) to use the whole plant or part of it as a herbal remedy e.g., cranberry, *Echinacea*, fever few, garlic, *Ginkgo biloba*, saw palmetto. The number of higher plant species (angiosperms and gymnosperms) on this planet is estimated to be 250,000 (Ayensu and DeFilipps, 1978), with a lower level at 215,000 (Cronquist, 1981; 1988) and at an upper level of 500,000 (Tippo and Stern, 1977; Schultes, 1972). Of these, only about 6% have been screened for biologic activity, and a reported 15% have been evaluated phytochemically (Verpoorte, 2000). Meanwhile, the need for basic scientific investigations on medicinal plants using indigenous medical systems has become imminent. In this Chapter, we analyze the past, present and future of medicinal plants, both as potential antibacterial crude drugs as well as a source for natural compounds that act as new anti-infection agents.

Documentation of Medicinal Plants in Asian Countries

Medicinal plants are best alternative remedies with tremendous opportunities. They not only provide access and affordable medicine to poor people; they can also generate income, employment and foreign exchange for developing countries. Many traditional healing herbs and plant parts have been shown to have medicinal value, especially in the rural areas and that these can be used to prevent, alleviate or cure several human diseases (Plate I). World Health Organization (WHO) estimates that more than 80% of the world's population rely either solely or largely on traditional remedies for health care. According to WHO a minimum of 20000 plant taxa have medicinal uses. It is estimated that up to 70000 plant species are used in folk medicine and a majority of these species are found in the Asia-Pacific region. However, the use of medicinal plants is faced with many constraints. Some of these constraints include: plants with medicinal values are not fully identified, inventoried and characterized, information and knowledge not being adequately documented and disseminated, many issues are not addressed and resolved (i.e. equity and sustainability), and the alarming commercial over-exploitation and consequent genetic erosion of medicinal plants.

Need for Conservation of Medicinal Plants

People who live in rural areas of the Asia-Pacific are familiar with the medicinal properties of plants, growing close to their homes, in the open fields, water margins, waste lands, both inside and outside of the forest areas and under different growth conditions. Most of the plant materials collected is used fresh either to obtain the extract from the whole plant or parts there of, whether they be leaves, roots, flowers or fruits. In case of woody forms, mostly the bark, roots and other parts are used. Carminatives like ginger, cloves and coriander are also usually added as fresh or dried materials. Though dried plant parts are frequently used, often the easy availability of fresh material is a critical point and the herbal doctor in the village

Ginger

Basak

Kalmegh

Aloevera

Tulsi

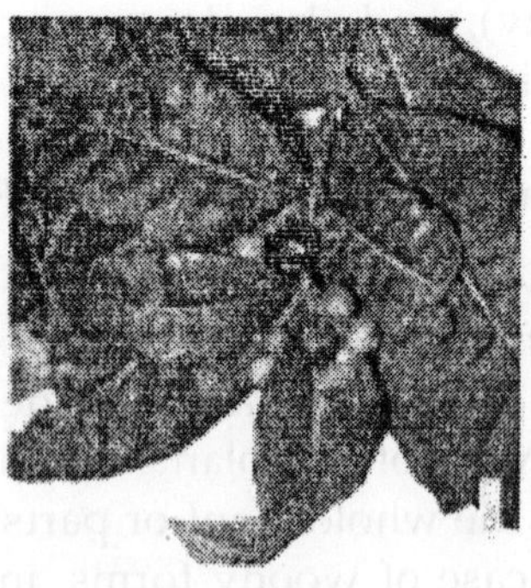

Rauvolfia

Thankuni

Moringa

Asparagus

Vangueria spinosa

Plate I: Some commonly used plants in India as traditional medicine

is well familiar with various plants he/she needs, their growth patterns, seasonality, habitat and other details. Such details were usually passed on in the past from parent to offspring in the family and uses of plants and the various combinations or mixes made were kept as a family secret. Along with the development of knowledge at family level, tremendous progress has been made at using the plant products at professional level in different societies, which have grown into branches of science in their own direction.

Though much information exists on the species diversity in medicinal plants in the Asia-Pacific region, relatively very little is known about the distribution, abundance, ecology and genetic diversity of the great majority of medicinal and aromatic plants, although some efforts have started in recent years (Gupta and Chadha 1995; Chandel et al., 1996; Kumar et al., 2000; Paisooksantivatana et al., 2001), including the use of molecular markers (Sharma et al., 2000; Natesh 2000). Identifying priority species for conservation and understanding the management requirements for most of these plants are constrained by the limited capacity for and attention to basic field research. The few thorough investigations of sustainable harvest conducted to date indicate that a combination of *in situ* protection of core populations, and controls on sustainable harvest involving local management regimes, must be combined with *ex situ* cultivation and conservation of rare and endangered, popular, and economically important species (Rao and Ramanatha 1998; Rao and Arora 2004). For the Indian Himalayan Region, a total of 1748 species of medicinal plants – 1020 herbs, 338 shrubs, 339 trees, apart from 51 pteridophytes – have been listed (Samant et al., 1998). These include several of the endangered medicinal plant species, using current IUCN, Red Data criteria under the Biodiversity Conservation Prioritization Project (BCPP), by Conservation Assessment and Management Plan (CAMP) workshop organized by WWF at Lucknow from 21-25 January 1997 (Samant et al., 1998).

Antibacterial Bioactive Phytocompounds from Extraction to Identification

Different approaches to drug discovery using higher plants can be distinguished: random selection followed by chemical screening; random selection followed by one or more biological assays; biological activity reports and ethnomedical use of plants (Eloff, 1998). The latter approach includes plants used in traditional medical systems; herbalism, folklore, and shamanism; and the use of databases. The objective is the targeted isolation of new bioactive phytocompounds. When an active extract has been identified, the first task to be taken is the identification of the bioactive phytocompounds, and this can mean either a full identification of a bioactive phytocompound after purification or partial identification to the level of a family of known compounds (Miles et al., 1998). For screening, plants are collected either randomly or by following leads supplied by local healers in geographical areas where the plants are found. Initial screening of plants for possible antimicrobial activities typically begins by using crude aqueous or alcohol extractions followed by various organic extraction methods (Rojas et al., 1992). Plant material can be used fresh or dried. Other relevant plant materials related to antimicrobial activities are the essential oils. Essential oils are complex natural mixtures of volatile secondary metabolites, isolated from plants by hydro or steam distillation and by expression (citrus peel oils).

Recio et al., (1989) reviewed the most relevant articles on this subject published between 1978 and 1988, compiling a list of 75 species in which the authors had established the activity of the extract along with both the spectrum of and the principles responsible for this activity.

In general, the review showed that phenolics are the predominant active chemical in these plants, with Gram positive bacteria being the most sensible germs. The review also revealed, however, the major problem with this type of research, namely the lack of uniformity in the criteria selected to study the activity. Many focus on determining the antimicrobial activity of plant extracts found in folk medicine

(Ngwendson et al., 2003; Bhattacharjee et al., 2006; Chatterjee et al., 2007; Ghosh et al., 2008), essential oils (Alma et al., 2003; Bhattacharjee et al., 2005) or isolated compounds such as alkaloids (Klausmeyer et al., 2004), flavonoids (Sohn et al., 2004), sesquiterpene lactones (Lin et al., 2003), diterpenes (El-Seedi et al., 2002), triterpenes (Katerere et al., 2003) or naphtoquinones (Machado et al., 2003), among others. Some of these compounds were isolated or obtained by bio-guided isolation after previously detecting antimicrobial activity on the part of the plant.

The main constituents of essential oils (mono and sesquiterpenes), along with carbohydrates, alcohols, ethers, aldehydes, and ketones, are responsible for the fragrant and biological properties of aromatic and medicinal plants. Due to these properties, since ancient times spices and herbs have been added to food, not only as flavoring agents but also as preservatives. Since nearly all of the identified components from plants active against microorganisms are aromatic or saturated organic compounds, they are most often obtained initially through ethanol and water extraction (Lewis et al., 1995). Some water-soluble compounds, such as polysaccharides like starch and polypeptides, including fabatin (Zhang and Lewis, 1997) and various lectins, are commonly more effective as inhibitors of virus adsorption and would not be identified in the screening techniques commonly used (Balzarini et al., 1991). Occasionally tannins and terpenoids may be found in the aqueous phase, but they are more often obtained by treatment with less polar solvents. An extraction to identification steps includes:

(1) Plant materials

(a) Plant can be chosen at random mainly based on previous literature or information collected on the basis of the local/ tribal peoples.

(b) Botanical identification of plant material as well as submission of Voucher specimen in the local herbarium.

(c) Climatic condition, geographical area and condition must be noted.

(2) Preparation of primary extract

(a) The plant material may be fresh or dried.

(b) Ground properly in order to optimize the solvent contact properly.

() Most of the literature revealed that primary extraction is mainly done with the help of ethanol/ methanol solvent as they have proved that ethanol/methanol extraction is more effective in isolating the bioactive phytocompound.

(3) Lyophilization

(a) After extraction lyophilization is generally recommended as this process is able to concentrate the material.

(b) The high concentration of the extract guarantees the identification of the bioactivity. Using low concentrations in drop tests may lead to false negative results.

(4) Secondary extract

(a) Sometimes due to complex composition of plant extract secondary extract is required which involves separation of micro molecules from macromolecules.

(b) The separation mainly involves the process of ethanol precipitation (30%v/v), centrifugation (10,000g for 10-15 min), and filtration by using different Millipore filters.

(c) Supernatant and precipitate are obtained and the main active ingredient (having antibacterial property) is generally found to be present in supernatant phase.

(5) Antibacterial screening

This is generally done by disk diffusion agar assay now commonly known as drop test.

(6) Determination of MIC

When antimicrobial activity is detected the minimum inhibitory concentration (MIC) must be determined to continue other antimicrobial assays of interest. The MIC is usually established by the broth dilution method.

(7) Bioautography guided chromatographic technique

(a) Bioguided chromatography techniques such as bioautography preceded by solvent separation is essential to initiate the bioactive phytocompound identification process; fraction collection with HPLC or FPLC assays, preparative TLC are also valid techniques. Bio-guided fraction and purification confirms previous results leading to isolation of a bioactive phytocompound.

(b) Separation of the compound by using solvents from non-polar to polar solvents.

(c) The following solvents are generally used to separate the ingredients in the following sequence (Mendonca -Filho, 2006):

(i) Cyclohexane

(ii) Hexane (Major identified compounds are: Essentail oil)

(iii) Toluene

(iv) Benzene

(v) Ether (Major identified compounds are: Alkaloids, Terpenoids, Coumarins, Fatty acids).

(vi) Ethyl acetate (Major identified compounds are: Flavonoids, Terpenoids)

(vii) Chloroform (Major identified compounds are: Steroids)

(viii) Acetone (Major identified compounds are: Flavonols)

(ix) Acetonitrile

(x) Butanol

(xi) 2- Propanol

(xii) Ethanol/Methanol (Major identified compounds are: Lectins, Alkaloids, Quassinoids, Flavones, Polyphenols, Tanins, Saponins).

(8) Bioactive fraction R_f and polarity determination

By TLC assays, R_f values can be determined and polarity of the groups can also be elucidated. Then by applying spraying reagent different phytocompounds can be identified as indicated in Table 1 (Ghosh and Chandra, 2006; Chowdhury et al., 2008a, 2008b; Ghosh et al., 2008).

(9) Determination of active ingredient

UV –vis, IR, NMR, HPLC/MS, GC/MS are commonly used to identify active ingredients.

Discussion

Medicinal plants constitute an effective source of both traditional and modern medicines, herbal medicine has been shown to have genuine utility and about 80% of rural population depends on it as primary health care. Over the years, the World Health Organization advocated that countries should interact with traditional medicine with a view to identifying and exploiting aspects that provide safe and effective remedies for ailments of both microbial and non-microbial origins (WHO 1978).

In the past few decades, the search for new anti-infection agents has occupied many research groups in the field of ethnopharmacology. Scientific criteria should be used in the selection of the plant material. In order to avoid use of random criteria, the selection of plants should be made from an ethnopharmacological perspective.

The solvent and the extraction system may both modify the final results (Nostro et al., 2000). The methodology employed is also important. For non-polar extracts, the use of diffusion techniques seems to be inadequate. Solid dilution techniques for studying plant extracts or non-polar compounds are most suitable. Diffusion technique is most suitable in case of less amount of sample. Choice of appropriate growth medium, and pH is another important criterion.

The potential for developing antimicrobials from higher plants appears rewarding, as it will lead to the development of a phytomedicine to act against microbes. Plant-based antimicrobials have enormous therapeutic potential as they can serve the purpose with lesser side effects that are often associated with synthetic antimicrobials (Iwu et al., 1999).

For the developing countries like India, the approval as drugs of standardized

Table 1

Detection of biochemicals present in TLC glass plate following treatment of different spraying reagents

Name of the Spraying Reagent	Result	Conclusion
Acetic anhydride-sulphuric acid (Liebermann-Burchard reagent)	Positive	Presence of steroids/ triterpene glycosides
Vanillin – phosphoric acid	Positive	Presence of steroids
Keller - Kiliani	Positive	Presence of cardiac glycosides
Anthrone reagent	Positive	Presence of ketose compound.
Antimony chloride	Positive	Presence of flavonoids
1% ethanolic solution of Aluminium Chloride	Positive	Presence of flavonoids
Ceric sulphate-sulphuric acid	Positive	Presence of steroidal glycosides
Dragendorff's reagent <Munier and Macheboeuf>	Positive	Presence of alkaloid substance
Mayor's solution	Positive	Presence of alkaloid substance
Formaldehyde-phosphoric acid	Positive	Steroid alkaloids / steroid sapogenins
Folin Ciocalteu reagent	Positive	Presence of phenolics
$FeCl_3$	Positive	Presence of phenolics
Silver nitrate	Positive	Presence of phenolics
Ninhydrin	Positive	Presence of aminoacids/amines/ amino sugars
Froth test	Positive	Presence of saponins
Bornträger's Reaction	Positive	Presence of Anthraquinones
p-Anisidine Hydrochloride	Positive	Presence of carbohydrates
Bromothymol blue	Positive	Presence of lipids and phospholipids
Diphenylamine	Positive	Presence of glycosides/glycolipids
m-Phenylenediamine	Positive	Presence of reducing sugars

and formulated plant extracts might be the starting point of an innovative and successful local pharmaceutical industry, which can compete with the large pharmaceutical companies, not only for the treatment of minor diseases, but also for the treatment of severe and life-threatening diseases. It can be stated that the major activities of natural products research of the past decades have clearly demonstrated that natural products represent an unparalleled reservoir of molecular diversity to drug discovery and development, and are complementary to combinatorial libraries.

The major disadvantage is the time taken to isolate and to characterize the active components from the extracts. By improving the diversity and quality of sample source and screen suitability, by accelerating dereplication and by automating and standardizing early isolation steps, the effectiveness of natural products research can be enhanced.

References

Alma, M.H., Mavi, A., Yildirim, A., Digrak, M. and Hirata, T. (2003). Screening chemical composition and *in vitro* antioxidant and antimicrobial activities of the essential oils from *Origanum syriacum* L. growing in Turkey. *Biological and Pharmaceutical Bulletin,* 26: 1725–1729.

Arevalo, G.V. (1994). Las plantas medicinales y su benificio en la salud: shipiboconibo. AIDESEP, Lima, Peru, p. 353.

Ayensu, E.S. and DeFilipps, R.A. (1978). Endangered and Threatened Plants of the United States. Smithsonian Institution and World Wildlife Fund, Washington, D.C. xv + 403 pp.

Balzarini, J., Schols, D.J., Neyts, E., Van Damme, W., Peumans., E. and Clercq, D. (1991). Alpha-(1-3)- and alpha-(1-6)-D-mannose-specific plant lectins are markedly inhibitory to human immunodeficiency virus and cytomegalovirus infections in vitro. *Antimicrobial Agents and Chemotherapy,* 35(3): 410–416.

Bhattacharjee, I., Chatterjee, S.K., Chatterjee, S.N. and Chandra, G. (2006). Antibacterial potentiality of *Argemone mexicana* solvent extracts against some pathogenic bacteria. *Memorias Instituto Oswaldo Cruz,* 101 (6):645-648.

Bhattacharjee, I., Ghosh, A. and Chandra, G. (2005). Antimicrobial activity of the essential oil of *Cestrum diurnum* (L.) (Solanales: Solanaceae). *African Journal of Biotechnology,* 2005; *4(4)*: 371 – 374.

Chandel, K.P.S., Shukla, G. and Sharma, N. (1996). Biodiversity in Medicinal and Aromatic Plants in India. National Bureau of Plant Genetic Resources, Pusa Campus, New Delhi, India, pp 235 – 239.

Chatterjee, S.K., Bhattacharjee, I. and Chandra, G. (2007). Bactericidal activities of some common herbs in India. *Pharmaceutical Biology,* 45(5): 350-354.

Chowdhury, N., Chatterjee, S.K., Laskar, S. and Chandra, G. (2008a). Larvicidal activity of *Solanum villosum* Mill (Solanaceae: Solanales) leaves to *Anopheles subpictus* Grassi (Diptera: Culicidae) with effect on non-target *Chironomus circumdatus* Kieffer (Diptera: Chironomidae). *Journal of Pest Science,* DOI 10.1007/s10340-008-0213-1.

Chowdhury, N., Ghosh, A. and Chandra, G. (2008b). Mosquito larvicidal activities of *Solanum villosum* berry extract against the dengue vector *Stegomyia aegypti BMC Complementary and Alternative Medicine,* 8:10 doi:10.1186/1472-6882-8-10.

Clavo, M., Seijas, Z., Alegre, J. (2003). Plantas medicinales usadas por mujeres nativas y mestizas en la region Ucayali.World Agroforestry Centre ICRAF, Pucallpa, Peru, p. 124.

Cronquist, A. (1981). An Integrated System of Classification of Flowering Plants. Columbia University Press, New York:, xviii, 1262 pp.

Cronquist, A. (1988). The Evolution and Classification of Flowering Plants. 2nd edition NY:New York Botanical Garden. Bronx, viii, 555 pp.

Desmarchelier, C., Schaus, F.W. (2000). Sixty medicinal plants from the Peruvian Amazon, 1st ed. Lima, Peru, p. 270 (privately published).

Eloff, J.N. (1998). Which extractant should be usedfor the screening and isolation of antimicrobial components from plants? *Journal of Ethnopharmacology*, 60: 1–8.

El-Seedi, H.R., Sata, N., Torssell, K.B. and Nishiyama, S. (2002). New labdene diterpenes from *Eupatorium glutinosum*. *Journal of Natural Products*, 65: 728–729.

Estrella, E. (1995). Plantas medicinales amazonicas: realidad y perspectivas. Tratado de Cooperacion Amazonica, Lima, Peru, p. 302.

Ghosh, A. and Chandra, G. (2006). Biocontrol efficacy of *Cestrum diurnum* (L.) (Solanales: Solanaceae) against the larval forms of *Anopheles stephensi*. *Natural Product Research*, 20(4):371–379.

Ghosh, A., Chowdhury, N. and Chandra, G. (2008). Laboratory evaluation of a phytosteroid compound of mature leaves of Day Jasmine (Solanaceae: Solanales) against larvae of *Culex quinquefasciatus* (Diptera: Culicidae) and nontarget organisms. *Parasitology Research*, 103: 271–277.

Ghosh, A., Das, B.K., Chatterjee, S.K. and Chandra, G. (2008). Antibacterial potentiality and phytochemical analysis of mature leaves of *Polyalthia longifolia* (Magnoliales: Annoaceae). *The South Pacific Journal of Natural Science*, 26: 68-72.

Gupta, R. and Chadha, K.L. (1995). Medicinal and Aromatic Plants in India.. In: *Advances in Horticulture: Medicinal and Aromatic Plants*. Vol. 11. (Eds., K.L. Chada,. and R. Gupta) Malhotra Publishing House, New Delhi. pp 1-44.

Iwu, M.W., Duncan, A.R. and Okunji, C.O. (1999). New antimicrobials of plant origin. In: *Perspectives on New Crops and New Uses* (ed. J. Janick), Alexandria, VA: ASHS Press, pp. 457-462.

Katerere, D.R., Gray, A.I., Nash, R.J. and Waigh, R.D. (2003). Antimicrobial activity of pentacyclic triterpenes isolated from African Combretaceae. *Phytochemistry*, 63: 81–88.

Klausmeyer, P., Chmurny, G.N., McCloud, T.G., Tucker, K.D. and Shoemaker, R.H. (2004). A novel antimicrobial indolizinium alkaloid from *Aniba panurensis*. *Journal of Natural Products*, 67: 1732– 1735.

Kumar, S., Hassan, S.A., Dwivedi, S., Kukreja, A.K., Sharma, A., Singh, A.K., Sharma, S. and Tewari, R. (eds). (2000). Proceedings of the National Seminar on the Frontiers of Research and Development in Medicinal Plants,16-18 September 2000. Journal Medicinal and Aromatic Plant Sciences Vol. 22/4A and Vol. 23/1A. Central Institute of Medicinal and Aromatic Plants (CIMAP). Lucknow, India.

Lewis, W.H. and Elvin-Lewis, M.P. (1995). Medicinal plants as sources of new therapeutics. *Annals of the Missouri Botanical Garden*, 82:16–24.

Lin, F., Hasegawa, M. and Kodama, O. (2003). Purification and identification of antimicrobial sesquiterpene lactones from yacon (*Smallanthus sonchifolius*) leaves. *Bioscience, Biotechnology and Biochemistry*, 67: 2154–2159.

Lossio, L.C., Delgado, H.S., Ruiz, A.G., Donayre, R.A., Ruiz, J.G., Panduro, M.P., Sifuentes, T.C., (1995). Plantas medicinales de la Amazonia Peruana. Instituto Peruano de Seguridad Social, Instituto de Medicina Tradicional, Iquitos, Peru, p. 255.

Machado, T.B., Pinto, A.V., Pinto, M.C., Leal, I.C., Silva, M.G., Amaral, A.C., Kuster, R.M. and Netto-dos Santos, K.R. (2003). *In vitro* activity of Brazilian medicinal plants, naturally occurring naphthoquinones and their analogues, against methicillin-resistant *Staphylococcus aureus. International Journal of Antimicrobial Agents*, 21: 279–284.

Mendonça-Filho, R.R. (2006). Bioactive Phytocompounds: New Approaches in the Phytosciences. *Modern Phytomedicine. Turning Medicinal Plants into Drugs* (Eds., I. Ahmad, F. Aqil, and M. Owais) WILEY-VCH Verlag GmbH & Co. KGaA, Weinheim ISBN: 3-527-31530-6. pp 1 – 24.

Miles, D.H., Nguyen, C.L. and Miles, D.H. (1998). Utilization of natural products for treatment of blood diseases. *Current Medicinal Chemistry*, 5: 421–440.

Natesh, S. (2000). Biotechnology in the conservation of medicinal and aromatic plants. In: *Biotechnology in Horticulture and Plantation Crops*. (Eds., K.L. Chadha, P.N. Ravindran and L. Sahajram), Malhotra Publishing House, New Delhi, India, pp. 548- 561

Ngwendson, J.N., Bedir, E., Efange, S.M., Okunji, C.O., Iwu, M.M., Schuster, B.G. and Khan, I.A. (2003). Constituents of *Peucedanum zenkeri* seeds and their antimicrobial effects. *Pharmazie*, 58: 587–589.

Nostro, A., Germano, M.P., D'Angelo, V., Marino, A. and Cannatelli, M.A. (2000). Extraction methods and bioautography for evaluation of medicinal plant antimicrobial activity. *Letters in Applied Microbiology*, 30: 379–384.

Paisooksantivatana, Y., Kako S. and Seko, H. (2001). Isozyme polymorphism in *Curcuma alismatifolia* Gagnep. (Zingiberaceae) populations from Thailand. *Scientia Horticulturae*, 88(4): 299-307.

Rao, A.N. and Ramanatha, R.V. (1998). Strategies for conservation of medicinal plants. *Symposium on Medicinal Plants: cure for the 21st Century (Biodiversity, Conservation and Utilization of Medicinal Plants)*, UPM, Serdang, Selangor, Malaysia.

Rao, V.R. and Arora, R.K. (2004). Rationale for conservation of Medicinal Plants. In: *Medicinal Plants Research in Asia* (Eds., P.A. Batugal, J. Kanniah, S.Y. Lee. and J.T. Oliver). Vol. I. The Framework and Project Workplans. International Plant Genetic Resources Institute – Regional Office for Asia, the Pacific and Oceania (IPGRI-APO), Serdang, Selangor DE, Malaysia, pp. 1-228.

Recio, M.C., R'ýos, J.L. and Villar, A. (1989). A review of some antimicrobial compounds isolated from medicinal plants reported in the literature 1978–88. *Phytotherapy Research*, 3: 117–125.

Reynel, C., Pennington, T.D., Pennington, R.T., Flores, C., Daza, A. (2003). Arboles utiles de la Amazonia Peruana y sus usos. Tarea Grafica Educativa, Lima, Peru, p. 250.

Rojas, A., Hernandez, L., Pereda–Miranda, R. and Mata, R. (1992). Screening for antimicrobial activity of crude drug extracts and pure natural products from Mexican medicinal plants. *Journal of Ethnopharmacology*, 35: 275–283.

Samant, S.S., Dhar, U. and Palni, L.M.S. (1998). Medicinal Plants of Indian Himalayas: Diversity, Distribution, Potential values. Himavikas Publication No. 13, G.B. Pant Institute of Himalayan Environment and Development, Almora, Uttaranchal, India. pp163.

Schultes, R.E. (1972). The future of plants as sources of new biodynamic compounds. In: *Plants in the Development of Modern Medicine* (Ed., T. Swain). Cambridge, MA: Harvard University Press, pp 103–124.

Schultes, R.E., Raffauaf, R.F. (1990). The Healing Forest. Dioscorides Press, Portland, USA, p. 484.

Sharma, K.D., Singh, B.M., Sharma, T.R., Katoch, M. and Guleria, S. (2000). Molecular analysis of variability in *Podophyllum hexandrum* Royle - An endangered medicinal herb of northwestern Himalaya. Plant Genetic Resources Newsletter, (FAO/IPGRI) 124: 57-61.

Sohn, H.Y., Son, K.H., Kwon, C.S., Kwon, G.S. and Kang, S.S. (2004). Antimicrobial and cytotoxic activity of 18 prenylated flavonoids isolated from medicinal plants: *Morus alba* L., *Morus mongolica* Schneider, *Broussnetia papyrifera* (L.) Vent, *Sophora flavescens* Ait and *Echinosophora koreensis* Nakai. *Phytomedicine*, 11: 666– 672.

Solecki, R. and Shanidar, I.V. (1975). A Neanderthal flower burial in northern Iraq. *Science* ,190: 880–881.

Tippo, O. and Stern, W.L. (1977). Humanistic Botany. New York:W.W. Norton.

Tshikalange, T.E., Meyer, J.J.M. and Hussein, A.A. (2005). Antimicrobial activity, toxicity and the isolation of a bioactive compound from plants used to treat sexually transmitted diseases. *Journal of Ethnopharmacology*, 96: 515-519.

Verpoorte, R. (2000). Pharmacognosy in the new millennium: lead finding and biotechnology. *Journal of Pharmacy and Pharmacology*, 52: 253–262.

World Health Organization (WHO) (1978). The promotion and development of traditional medicine. *Technical report series*, pp. 622.

Zhang, Y. and Lewis, K. (1997). *FEMS Microbiology Letters*, 149: 59–64.

5

Biological Control of Insect Pests by their Natural Enemies

B.M. Sharma

Introduction

Although insects are in an integral part of the ecological web, and insects pests are responsible for 25-30% loss of the world's food production. The losses due to insect pests to fodder and forest trees are quite accountable in agroforestry (Sharma, 2006). Our dependence of pesticides have increased upto the extent that if modern agriculture will attempt to operate without them, crop production will decline, food prices will soar far higher and food shortage will become even more severe. However, pesticides have created two major serious problems: firstly, many of them have persisted and accumulated in the environment and contaminated the non-target organisms and, secondly, many of them have directly or indirectly affected human health. In opinion of US National Academy of Science the natural soil micro and macro-organisms should be used to increase crop fertility rather than the use of insecticides. The committee has also recommended the formation of new generation of biological agents that can destroy the target insects. The use of pesticides have been reckoned as double edged swords of ecological narcotics much because of their negative potentials. Forest ecosystems have much less insect pest problems than crop fields, because these are a large number of natural enemies. They are most numerous and diverse against insect pests (Sharma, 2008).

India, one of the world's largest agricultural economics, poses no exception. Demand for chemical pest control and resulting negative externalities have explanted greatly during the last four decades (Chand and Birthal, 1997). The population dynamics of insect pests in agroecosystems may often be linked to those of other phytophagous species through the foraging activities of natural enemies; these indirect interactions provide both opportunities and challenges for biological control. Biological control is most successful in temperate region comparison to tropical and subtropical regions. Biological control is the use of living organisms to suppress pest populations, making them less damaging than they would otherwise be. It can be used against all types of pests, including vertebrates, plant pathogens and weeds as well as insects.

One of the total insect species feed upon insects either as predator or parasitoids keeping down their population growth (Pandey *et al.*, 2004). Although most serious pests are exotic species from abroad, indigenous natural enemies can be used to control them if they are released by seasonal inoculative release. This procedure is preferable to the use of imported species, from the point of view of the possible effect of imported species on indigenous species in agroecosystems. Promising results were obtained after the use of biological agents against insect pests.

Micro- and macro-organisms degrade pesticides residues to less toxic compounds. Pesticide degradation by enzymes is further attractive. Immobilized enzymes are best for pesticide degradation because the transport of pesticides into microbial cell is difficult and enzymes can be lost in one attempt. Pesticides degradation characters have been located on plasmids, which provide clues for genetic manipulation for development of new microbial strains capable of efficiently degrading pesticides. The increasing cost of plant protection and accelerating pest incidents make agriculture a risky and less profitable enterprise (Chopra, 1993). At the same time the toxic materials generated from chemical farming pollute the environment and harm consumers' and farmers' health. A more economical and eco-friendly alternative is the adoption of biological control. Furthermore, the biological agents not only harmless, more effective, self perpetuating, ecologically sound, economical and free from environmental pollution, toxic health hazards and other undesirable side effects.

Historical Development

Since ancient times, Horticulturist of Asian countries is using predaceous ants for reducing the number of foliage feeding insects. Erashus Darvin (1800) suggested in his book Phytologia that the predators and parasites may control biologically as an ichneumon fly killing eggs of butterfly caterpillar. After then many scientists think in this direction and made successful experiments. The remarkable success in this direction was the beetle (*Rodolia cardinalis*) from Australia in California (U.S.A.) in 1888. This lady bird beetle was used in 32 countries of the world where it established well and completely checked the population of cushion scale (*Iceya purchasi* Mash).

Georgh Campier (1890) made survey of different countries of the world and collected predators and parasites of fruits and sent to America. Mure (1906) surveyed the natural enemies of sugarcane pests, and found *Scolia moniliae* was successfully controlling the *Anomala orientales*. The term biological control was first used by Smith (1919) to signify the use of natural enemies to control insert pests. Many *Trichogramma* spp. Have been observed parasitizing the eggs of insect pests of several vegetable and forest plant species (Ahmad, 1990). It has assumed this status owing to its maximum exploitation as biocontrol agent against many insect species. They are small in size, short life cycle, relatively easy handling and rearing, adaptability to varied environmental conditions as well as compatibility with the host species. Evans and England (1996) conducted field experiments to explore indirect interactions in the biological control of the alfalfa weevil (*Hypera postica*). Pea aphids (*Acyrthosiphon pisum*) produced honeydew in alfalfa fields may be important as a source of nutrition for adults of the wasp *Bathyplectes curculionis* and may thereby enhance parasitism of the wasp's host, the alfalfa weevil. The presence of aphids may also promote aggregation of lady beetles, which consume weevil larvae in alfalfa. But by feeding heavily on pea aphids as well, these predators could indirectly enhance weevil survivorship by undercutting the wasp's capacity to parasitize its host. Some scientists also believe that the most promising weapons by seeing their effectiveness in nature so that some natural enemies are reared in the laboratories and utilized for the biological control of insect pests. (Swan, 1964; Singh, 1994; Vidyasekaran *et al.*, 1992).

Insect viruses are attractive as biological control agents and could be a feasible alternative to chemical insecticides in the management of insect infestations. Recent advances in the development of wild-type and genetically modified viruses as insecticides. The strategy of application of insect viruses and assessment of bio-safety of genetically modified *Helicoverpa armigera* Nucleopolyhedovirus (HearNPV) is reviewed by Xiu-lian and Hui-yin (2007).

The invasion of *Parthenium* was at alarming rate during the year 2000 in uttarakhand state but after attack by Maxican beetle it observed that this weed suppressed after some time (Goyal and Brahma, 2001). Sushil Kumar (2008) reported *Parthenium* infestation in about 10 million hactares of land based on the extensive survey throughout India. The pathogens and mycoherbicides against weeds has been reviewed by Evans, 1997, Kumar, 1998; Pandey and Saini, 2002; Bisht, 2004; and Oudhia, 2005.

Macro-organisms Used for Biological Control

Predatory and Parasitic Insects

More than 200 families in 15 orders have been recorded as entomophagous insects. A total 403 spp. of natural enemies were introduced from other countries in which only 94 established in the country. A large number of predators and parasites have been introduced from different parts of the world in India for the biological control of harmful insects and only 13 are successfully utilized.

Parasites: It is an organism which is usually much smaller than its host and a single individual generally does not kill the host. Parasite may complete their entire life cycle on a single host (e.g., lice), or they may be free living and non parasitic during part of their life cycle (mosquitoes and fleas) or they may have complicated life cycles involving several host species (tapeworms).

Parasitoids: A parasitoid is a special kind of predator which is often the same size as its host, kills its host and requires only one host (prey) for development into a free

living adult. Some common parasitoids are given below which are being used for the control of insect pests. *Trichogramma* is exclusively an egg parasitoid of over 220 insect species belonging to the orders Lepidoptera, Coleoptera, Neuroptera, Diptera and Hymenoptera. It is widely distributed in all six zoogeographical regions (Ahmad *et al.*, 2008).

S.No.	Parsitoid	Host Insect
1.	*Phytoseiulus persimilis, Amblysius californicus, A. cucumeris, Feltiella acarisuga, Stethorus punctillum* and *Macrolophus caluginosus*	Spider mites
2.	*Encarsia formosa* and *Eretmocerus* spp.	White flies
3.	*Typhlodromips swirskii*	Spider mites, white flies and thrips
4.	*Trichogramma* spp.	Lepidopterans, Coleopterans, Neuropterans, Dipterans and Hymenopterans

Predators: A predator is a free living organism throughout its life, it kills the prey and is usually larger than prey and requires more than one prey to complete development. Predators are those insects which catch and devour the harmful insects while parasite makes their homes on or inside the body of the host and pass their life cycle and finally destroy the host insect. Some of the predators and parasites are given below which have been successfully used in different parts of the world of the biological control insects.

S. No.	Imported predator/ parasite	Host insect
1.	*Rodolia cardinalis* (predator)	Cottony Cushion Scale (*Icerya purchasis*)
2.	*Cytorhinus mundulus*	Sugarcane Leaf Hopper (*Perkinsiella saccharicida*)
3.	*Coccophagus gurneyi* (parasite)	*Cirtophilus mealybug*
4.	*Tetranemus pretisus*	*Pseudococcus gahani*
5.	*Anarhopus sydneyensis*	*Long-tailed mealybug*
6.	*Tetranemus perigrinus*	*Pseudococcon adonidum*
7.	*Pseudopycust utilis*	*Pseudococcus nipae*
8.	*Microceromasia sphcnophori*	Sugarcane weevil (*Rhabdoenemis obscura*)

Natural Enemies Imported in India

The following 13 natural enemies are found successful which have been imported from different parts of the world in India to control insect pests.

S. No.	Imported predator/ parasite	Host insect/ vertebrate/weed
1.	*Rodolia cardinalis* (predator)	Cottony Cushion Scale (*Icerya purchasis*)
2.	*Euglandina rosea*	Giant african snail (*Achetina fulica*)
3.	*Aphelinus mali* (parasite)	Wolly Aphid (*Eriosoma lanigerum*)
4.	*Bracon gelechae*	Potato Tuber Moth (*Gnorimochema operculla*)
5.	*Prospaltella perniciosi*	Sanjose Scale (*Quadraspidiotus perniciosus*)
6.	*Aphytis mytilaspidis*	
7.	*Spoggosia beniziana*	Coconut Caterpillar (*Nephantis serinopa*)
8.	*Opius vandenboschi*	Fruit fly (*Dacus* spp.)
9.	*Apanteles flavipes*	Jower Stem Borer (*Chilo partellus*)
10.	*Telenomus nawai*	Castor Semilooper (*Achoea janata*)
11.	*Trichogramma minutum*	Sugarcane and Cotton Borers
12.	*Lixophaga diatraeae*	Sugarcane Borers
13.	*Syngamia haemorrhoidalis*	Lantana Weed

Vertebrate Predators

Different types of vertebrates found to catch and devour the insects viz. fishes, frog, toad, tortoise, snakes, lizard, birds (Hudhud, Maina, Bule-Jay, Partridge, Quail, Peacock, Papiha, Owl, Bhujanga, Cuckoo, Crow, Sparrow and Kite), mammals (Bat, Rat, Mole and Mongoose). However, among the predatory vertebrates reptiles and birds are most important but they are less used to introduce from one country to another. These natural enemies are doing as an autocidal control of insects in the nature so that they are being used as biological control agents.

Micro-organism Used for Biological Control

Very small and microscopic organisms are also being used as biological agent. They degrade pesticidal and hazardous residues to less toxic compounds.

Bacteria

De Herelle (1916) gave the first practical biological control of insects on the use of bacterium *Coccobacillus acridiorum* against locusts. In the beginning of 1940 in the United States, it was observed that milky disease A and B in Japanese beetle (*Popillia japonica*) is caused by the bacteria *Bicillus popillae* and *Bacillus lentimorbus*. These bacteria are successfully utilized for the control of Japanese beetle grubs. Infected larvae are crushed and mixed into the soil where they parasitise and kill the larvae. Now days, *Bacilius thuringiensis* is successfully used for the control of lepidopterans pests in different parts of the world. The various commercial formulations are available which are used to control the pests e.g., Dipel, Thuricide, Biolip, Biotox etc. The spore forming bacteria are more promising in insect control because they form endospores which are highly resistant to environmental changes. Amongst the spore formers, the crystalliferous are better than non-crystalliferous because they produce crystals of toxic nature (Upadhyay and Mathur, 1998).

Fungi

The fungi are being used for the control of injurious insects. These attack the host body externally or internally where the spores germinate and cover the whole body and kill the host. In America, White Muscardine fungus *Beauveria bassiana* has been successfully used for the first time, it used against chinch bug (*Blessus lucupterous*). Spare and Colley (1912) used the fungus *Entomophthora aulicae* for the successful control of brown tail moth. *Empusa* species has also been used against houseflies and grasshoppers. All the spores of above fungi are mixed in water and sprayed on the crop. These fungal spores attach to body of host where they germinate and produce mycelium and hyphae and kill the host. High humidity is needed for the successful germination of the spores. Aflatoxins, destruxins and some other unidentified toxins are produced by entomogenous fungi. The alfatoxins cause changes in the metabolic activity of the insects while destruxins (A & B) cause paralysis and death in insects.

Species of genus *Trichoderma* are used to manage some soil-borne plant pathogens. *Beuveria bassiana* is used to manage different types of pest such whiteflies, thrips, aphids and weevils. *Paecilomyces fumosoroseus* is also used for different pests such as white flies, thrips and aphids. *Metarhizium anisopliae* is used against spider mites. Other fungi which are effective for the biological control of insects are *Isaria, Spororichum, Anisoplite, Aspergillus* spp. etc.

Viruses

Although the important role of viruses as natural control factor have long been recognized. But, it is only in the last few years that the viruses have been used successfully in control programme. First successful establishment of an exotic virus disease was accomplished by Balch (1946), who introduced the Polyhedrosis virus of the European spruce-saw fly (*Diprion heraynial*) into New found land. The first successful attempt to introduce a virus which was not present previously in the North American continent was made by Bird (1953) who applied a Polyhedrosis virus of the European pine saw fly (*Neodiprion sertifer*) obtained from Sweden. Now days, Polyhedrosis virus (*Borrelina campeolis*) has been successfully utilized in California for the biological control of Alfalfa by spraying of virus suspension containing 50 lac Polyhedra/c.c.@5 gallon/acre. Steinhaus (1949) also used the Polyhedrosis virus successfully against *Heliothis armigera* and *Heliothis virescens*.

The inclusions of viruses are more promising as they produce inclusion bodies in the host cell. The Polyhedrosis virus produces polyhedral bodies (many-sided bodies) and they are called as Nuclear Polyhedrosis Virus (NPV) or the cytoplasm than called as Cytoplamic Polyhedrosis Virus (CPV). In India NPV of *Heliothis armigera* is being used on field scale @ 250 LE/ha (larval equivalent) to control gram pod borer.

Protozoan

There are so many parasitic protozoans as reported by Tanada (1959) that the parasite protozoan may be used successfully as a biological control agent because of their abundance. But due to various hindrances in their mass production not much work has been done on this aspect. It may be one of the causes that they produce chronic disease infection in spite of acute disease as a result of which the host insects may be survive for a longer period. In America, *Perezia pyraustia* has been successfully utilized for the control of European corn borer (*Ostrinia nobilalis*).

Nematodes

Some nematode species are also known which parasitize and destroy the insects. Glacer and Wilaox (1918) described in detail for using nematodes as a biological agents. Tanada (1959) also reported the presence of nematodes as international parasite on various beetles, grasshoppers, moths and other insects recommended that they may be used successfully for the control of these insects. Currently, DD-136 (*Neoaplectana carpocapsae*) is being used against the control of various insect pests. The nematode suspension is sprayed on the crops for the control of lepidopterous, coleopterous and other insect species. In India, it has been found successful against a number of pests. The other nematode species are *Neotylenchus, Rhabditis, Mermis, Hercamermis, Protrellus, Pseudonymus* etc. The entomopathogenic nematodes (EPNs) can cause quick mortality in the major insect pests of crops through infection by associated bacterial pathogens. These are amenable for mass production and inundative applications (David, 2006).

Characteristics of a Natural Enemy

The most effective character of a natural enemy is searching capacity of their host. When the host population is scarce it should be able to find and dispersal the host. It should be able to control such pest for which is being introduced and never should be phytophagous. It should be highly specific of monophagy but do not pass their life on other hosts. It should be high fecundity rate and short life cycle, so that they may be produced in large number in a short period. More generations of the enemies may be developed to the generation of the host. It should be adoptable to new environmental conditions and resistant with the pesticides.

Merits of Biological Control

Uses of natural enemies are basic requirement to stop pesticidal pollution and economic loss. These are so toxic but they do not leave any harmful residues and do not harm to beneficial organisms as like chemical control. The effect of biological control is permanent type when it once introduced may work constantly in the nature for several years. Some agents can easily be mass produced economically and may be utilized. It is free from resistant problems as like insecticidal resistance. Some agents may be used in combination with insecticides. If insect save from the attack of insecticide may be controlled by the natural enemies.

Demerits of Biological Control

Since the number of natural enemies available for biological control is limited. These biological control agents have narrow host range and control only a particular pest. The effect is very slow and required a long time for the control. Highly trained personals are required to use so that much money and time required. More members of biological control agents are required. After some times introduced agents are also parasitized by another parasite. Due to lack of suitable environmental conditions the introduced parasites or predators do not adopt the prevailing conditions. More time is required to study the adoptability of that agent.

Future Prospectus

To maintain a pesticide free healthy environment and obtain a sustainable supply of natural agricultural products, it should be carefully managed and conserved at any cost. The necessary long-term empirical research and commitment of scientists working in developing countries cannot be evaluated solely by the number of publications. Fundamentally, new and closer partnerships are needed between farmers, entomologists, extension and research workers, and agricultural development agencies to disseminate information and impart training in the field of biological control.

References

Ahmad, M. (1990). Potentiality of egg parasites, *Trichogramma* spp. Against defoliating pest of teak, *Tectona grandis L.F. Ann. Entomol.*, 8(1): 19-22.

Ahmad, M., Ahmad, M.J. and Mishra, R. K. (2008). Biosystematics and functional diversity of *Trichogramma*. In Pests of Forest Importance and their Management (Eds., B.K. Tyagi, V. Veer and S. Prakash), pp. 299. Scientific Publishers, Jodhpur, India.

Bisht, G. (2004). Experts turn to beetles to check the growth of congress grass. *Indian Express*, June 4, 2004.

Chand, R. and Birthal, P.S. (1997). Pesticide use in Indian agriculture in relation to growth in area and production and technological change. *Indian Journal of Agricultural Economics*, 52(3).

Chopra, K. (1993). Sustainability of agriculture. *Ind. J. of Agricul. Eco.*, 48(3): 527-537.

David, B.V. (2006). Eelworms (nematodes) as both targets and tools for biocontrol in indian agriculture. Current Sc., 90(5): 630.

Evans, E.W. and England, S. (1996). Indirect Interactions in Biological Control of Insects: Pests and Natural Enemies in Alfalfa. *Ecological Applications*, 6 (3) : 920-30.

Evans, H.C. (1997). *Parthenium Hysterphorous*: A review of its weed status and possibilities of biological control. *Biocontrol News Information*, 18: 89-98.

Goyal, C.P. and Brahma, B.C. (2001). A Ray of Hope against *Parthenium* in Rajaji National Park. *Indian Forester*, 127(4): 409-14.

Oudhia, P. (2005). Strategy for *Parthenium* management from the *Parthenium* action group- Indian Experience. *Proc. 2nd Intel. Conf. on Parthenium management*, Bangalore, pp. 215-218.

Pandey, P., Kumar, N. and Tripathi, C.P.M. (2004). Impact of males on the progeny sex ratio of *Campoletis chlorideae* (Hymenoptera: Ichneumonidae), a parasitoid of *Helicoverpa armigera* (Hübner) (Lepidoptera: Noctuidae). *Journal of Applied Entomology*, 128: 254-257.

Pandey, P.K. and Saini, S.K. (2002). *Parthenium* threat to biological diversity of natural forest ecosystem in Madhya Pradesh. *Vaniki Sandesh*, 26:21-29.

Sharma, B.M. (2006). Deforestation: A man made disaster. Science, India.

Sharma, B.M. (2008). Impact of man, environment and pest on forest resources. In: *Forest Insect Pests and Control* (Eds., B.K. Tyagi, V. Veer and S. Prakash), Scientific Publishers, Jodhpur, India.

Singh, S.P. (1994). Fifteen years of AICRP on Biological control. Parishree Printers, Bangalore, p. 320.

Smith, H. S. (1919). On some phases of insect control by the biological method. *J. Econ. Ent.*, 12 : 288-92.

Steinhaus, E. A. (1949). *Principles of Insect Pathology*. McGraw-Hill Book Co., Inc., New York. 757p.

Swan, L.A. (1964). *Beneficial Insects*, p. 249.

Upadhyay, K.D. and Mathur, V.K. (1998). Insect Pests, Rastogi Publication, Meerut.

Vidyasekaran, P., Balasubramanium, G. and Sundra Babu, P.C. (1992). Technological advances in integrated pest management for sustainable agriculture. In: *Sustainable Agriculture*. Tamil Nadu Agriculture University Press, pp. 55-73.

Xiu-lian, Sun and Hui-yin, Peng (2007). Recent advances in biological control of pest insects by using viruses in China *Virologica sinica*, 22(2):158-162.

6

Natural Radioactivity in Environment

Dmitro M. Grodzinsky

Introduction

All biotic and abiotic components of environment everywhere in Nature are radioactive because they store natural radioactive elements. Over 60 natural radioactive isotopes, radionuclides are known. They are classified in three groups:

1. Primordial radionuclides which are contained in the materials of our planets since the creation of the Earth.
2. Cosmogenic radionuclides which are forming as a result of the cosmic rays interactions with atoms of some chemical elements.
3. Artificial radionuclides which arise in the nuclear reactions due to human activities in the field of nuclear technologies and researches.

The radionuclides involved in mentioned groups can be found everywhere in air, continental and ocean water, rocks, soil, plants, animals, fungus and microorganisms. Radionuclides are extended everywhere in the dispersed condition when the main interactions between different substances and radionuclides are due to adsorption processes. Being concentrated in different objects of environment radionuclides stipulates a permanent irradiation of all living beings with ionizing radiation. Annual estimated average dose rate in Biosphere is equal to some tenth of cG rays per year being varied in a broad range according to local concentration of radionuclides in the objects of environment.

In practice there are no places and objects on our planet with zero radio activities. Every living creature contains significant quantities of Carbon-14 (^{14}C) and Potassium-40 (^{40}K), which are the main radionuclides as isotopes of chemical elements incorporated into any cells of organism. The levels of dose-rate of irradiation is very low in the most cases because contains of the natural radioactivity is not too high. However there are two situations when the dose-rate of irradiation can be much higher in comparison with the average level of irradiation. The first one is related to territories characterized by high concentration of radionuclides in soil or rocks. The second on is due to radioactive contamination of environment with artificial technogenic radionuclides. High level of radioactive contamination may pose major risks to people as well as to environment. Different species can be exposed to very high levels of irradiation from the spread of contamination following nuclear accidents or other paths of radionuclide injection into environment. Hence the content of radionuclides in soils rocks, water and air should be very important parameters of the ecological quality of environment (Danil'chenko and Grodzinsky, 2004).

Units Used in Description of Radionuclide Dispersion in Environment and Doses of Irradiation

Surface dispersion of radionuclides is usually expressed in units of radioactivity per unit of area. In System International, SI a unit of radioactivity is Becquerel (rate of radioactive decay is equal to 1 decay per second) and surface contamination is expressed in Becquerel per square meter, Bq/m^2. Sometimes off-system prior unit Curie (Ci) is used. 1 Ci is equal to 3.7×10^{10} Bq and then dimensions of unit for surface dispersion of radionuclide a[illegible] Curie per square kilometer, Ci/km^2. Volumetric content of radinuclide is expressed in unit of specific radioactivity - radioactivity per unit of volume or weight. This value has a dimension Bq/L or Bq/ kg or Ci/L, C/kg etc. Many of the units are broken out into smaller units or expressed as multiples, using standard metric prefixes.

Doses of irradiation are expressed in Gray (Gy) who is equal to 100 rads. A rad is a unit of absorbed dose of ionizing radiation equal to energy of 100 ergs per gram of irradiated material. Accordingly Gray is equal to energy of one joule per one kg of irradiated material. Gy is a unit of absorbed dose irradiation. The equivalent effective dosage of ionizing radiation is expressed in other units. For example, an off-system unit rem (roentgen equivalent man) is a dosage of an ionizing radiation that will cause the same biological effect as one roentgen of X-rays or gamma ray dosage. Roentgen is very old off-system unit of X-radiation or gamma radiation. It is equal to the amount of radiation that produces in one cubic centimeter of dry air in 0^0 C and standard atmospheric pressure ionization of either sign equal to one electrostatic unit of charge. The Sievert (Sv) is a unit of effective equivalent dose of any tapes of radiation. It is equal to the amount radiation that will cause the same biological effect as one Gray. Gy and Sv are concerned to harmful doses of irradiation. Lethal dose for acute irradiation of human is about 5 Sv. Units of dose-rate are as follows: gray per hour (Gy/hr), Sv per hour (Sv/hr), rem per hour (rem/hr) grays per year (Gy/yr) and so forth. All of the preceding pertains equally to a dispersion of the natural radioactivity as well as to contamination of the environment with the artificial radionuclides.

Primordial Radionuclides

Primordial radionuclides are components of chemical composition from when the world and Universe have been created. These radionuclides are typically long lived and half-lives for these isotopes amount often on the order of hundreds of millions of years. There are two groups of primordial radionuclides: a group of heavy radioactive elements related to the last numbers of the periodic tables of the chemical elements and a group of more light radioactive elements. The heavy primordial radioactive elements make up three ancestors of chain radionuclides, which are formed as results of successive radioactive decays. The products of radioactive decay of radioisotopes relating to the heavy radioactive elements are also radioactive nuclides, which undergo radioactive decay forming a chain of radioactive conversion. These chain processes are lasting till arising stable isotopes. The series of radioactive transformations in chain are named as radioactive families. The first radioisotope, which begins a series of radioactive transformation, is an ancestor or parental nuclide of radioactive family. There are three parental radionuclides and correspondingly three radioactive families. They are Thorium - 232 family, Uranium - 235 family and Uranium - 238 family. There are plentiful radionuclides for every parental radionuclide in its decay chains (Grodzinsky, 1965). The main ancestors of heavy primordial radionuclides are shown in Table 1.

Table 1

Ancestors of heavy primordial radionuclides

Nuclide	Symbol	Half-life, year	Type of radioactive Decay Radiation	Energy of particles or gamma-quantum, Mev	Percent Yield
Uranium - 235	^{235}U	7.1 10^{8}	α	4.32	3
				4,21	3.7
				4.58	8
				4.50	1.2
				4.40	57
				4.37	18
			γ	0.110	2.5
				0.143	11.0
				0.163	5.0
				0.185	54.0
				0.205	5.0
Uranium-238	^{238}U	4.5 x 10^{9}	α	4.20	75
				4.15	23
			γ	0.048	23
Thorium- 232	^{232}Th	1.4 x 10^{10}	α	4.01	76
				3.95	24
			γ	0.55	24

The main characteristics of all radionuclides related to series of heavy radioactive isotopes are listed in Tables 2 - 4

Table 2
Nuclides of the ^{238}U Series

Nuclide	Relative isotopic Abundance, %	Half-Life	Type of radiation	Energy, MeV	Percent yield
^{238}U	99.28	4.5 x 10^9 yr	α	4.30	75
			α	4.15	23
			γ	0.048	23
^{234}Th	-	24 days	$β^-$	0.192	65
			$β^-$	0.100	35
			γ	0.092	4.0 (doublet)
^{234m}Pa	-	1.2 min	$β^-$	2.29	98
				1.53	< 1
				1.25	< 1
			γ (IT)	0.39	0.13
			γ	0.817	4
^{234}U	0.0057	2.5 x 10^5 yr	α	4.77	72
				4.72	28
			γ	0.093	
^{230}Th	-	8.0 x 10^4 yr	α	4.68	76
				4.61	24
				4.51	0.35
			γ	0.068	0.6
				0.253	0.02
^{226}Ra	-	1622 yr	α	4.78	94.3
				4.59	5.7
			γ	0.186	4
				0.26	0.007
^{222}Rn	-	3.8 days	α	5.48	100
			γ	0.510	0.007
^{218}Po	-	3.05 min	α	6.0	99
^{214}Pb	-	26.8 min	$β^-$	0.72	100
			γ	0.053	1.6
				0.242	4
				0.295	10
				0.352	36
^{218}At	-	1.5 - 2 sec	α	6.70	94
				6.65	6
^{214}Bi	-	19.7 min	$β^-$	3.26	19
				1.51	40
				1.00	23
				1.88	9
			α	5.52	0.008
				5.45	0.011
				5.27	0.001

^{214}Po	-	1.64×10^{-4}	α γ	7.68 0.799	100 0.014
^{210}Tl	-	1.3 min	β^- γ	1.9 1.3 2.3 0.296 0.795 1.31	56 25 19 80 100 21
^{210}Pb	-	22 yr	β^- γ	0.015 0.061 0.0465	81 19 4
^{210}Bi	-	5.0 days	β^- α	1.17 5.0	99 5×10^{-5}
^{210}Po	-	138 days	α γ	5.3 0.80	100 0.0011
^{206}Tl	-	4.2 min	β^-	1.51	100
^{206}Pb	25.2	Stable	-	-	-

Table 3
Nuclides of the ^{235}U

Nuclide	Relative nuclide abundance, %	Half-Life	Type of radiation	Energy, Mev	Percent yield
^{235}U	0.72	7.1×10^{8} yr	α	4.32	3
				4.21	5.7
				4.58	8 (doublet)
				4.5	1.2
				4.4	57
				4.37	18
			γ	0.110	2.5
				0.143	11.0
				0.163	5.0
				0.185	54
				0.205	5.0
		25.64 hr	β^-	0.302	52
				0.218	20
				0.138	22
			γ	0.026	2
				0.085	10
^{231}Pa	-	3.25×10^{4} yr	α	5.00	24
				4.94	22
				5.02	23
				5.05	10
			γ	0.027	6
				0.29	6 (complex)
^{227}Ac	-	21.6 yr	β^-	0.046	99
			α	4.95	1.2 (doublet)
			γ	0.070	0.08
^{227}Th	-	18.2 days	α	6.04	
				5.98	23
				5.76	24
				5.72	21
					14 (doublet)
			γ	0.050	
				0.237	8
					15 (complex)
				0.31	8 (complex)

^{223}Fr	-	22 min	β^- α γ	1.15 5.35 0.050 0.080 0.234	99 4×10^{-3} 40 13 4
^{223}Ra	-	11.4 days	α γ	5.75 5.71 5.61 5.54 0.149 0.270 0.33	9 54 26 9 10 (complex) 10 6 (complex)
^{219}Rn	-	4.0 s	α γ	6.82 6.55 6.42 0.272 0.401	81 11 8 9 5
^{215}Po	-	1.77×10^{-3} s	α β^-	7.38 -	99 2.3×10^{-4}
^{211}Pb	-	36.1 min	β^- γ	1.36 0.95 0.53 0.405 0.427 0.832	92 1.4 6 3.4 1.8 3.4
^{215}At	-	10^{-4} s	α	8.00	100
^{211}Bi	-	2.16 s	α β^- γ	6.62 6.28 - 0.35	84 16 0.27 14
^{211}Po	-	0.52 s	α γ	7.45 6.89 0.57 0.90	90 0.5 0.5 0.5
^{207}Tl	-	4.79 min	β^- γ	1.44 0.870	100 0.16
^{207}Pb	21.7	Stable	-	-	-

Table 4
Nuclides of the Thorium Series

Nuclide	Relative Nuclide abundance, %	Half-life	Type of Radiation	Energy, Mev	Percent Yield
^{232}Th	100	1.4×10^{10} yr	α γ	4.01 3.95 0.055	76 24 24
^{228}Ra	-	6.7 yr	β^-	0.055	100
^{228}Ac	-	6.13 hr	β^- γ	2.18 1.85 1.72 1.11 0.64 0.46 0.058 0.129 0.184	10.0 9.6 6.7 53.0 7.6 13 53.0 5.9 < 1
^{228}Th	-	1.9 yr	α γ	5.42 5.34 0.083	71 28 1.6
^{224}Ra	-	3.64 days	α γ	5.68 5.45 5.19 0.241	94 6 0.4 3.7
^{220}Rn	-	55 s	α γ	6.28 0.50	100 0.07
^{216}Po	-	0.16 s	α	6.77	100
^{212}Pb	-	10.6 hr	β^-	0.33 0.57 0.176 0.238 0.300	88 12 < 1 47 3.2

^{212}Bi	-	60.5 min	β^- α γ with β γ with α	2.25 6.086 6.047 1.81 1.61 1.03 0.83 0.72 0.040 0.288 0.46	64 9.8 25.1 1.0 1.8 2 13 7 2 0.5 0.8 (complex)
^{212}Po	-	3.04×10^{-7} s	α	10.55 8.785	< 1 99
^{208}Tl	-	3.1 min	β^- γ	1.80 2.61 0.86 0.58 0.51	10 100 12 86 23
^{208}Pb	52.3	Stable	-	-	-

The linkage between different members of radioactive series (families) is well seen in the schemas of radioactive decay. The next schemas characterize series of radioactive transformation in three radioactive series (radioactive families).

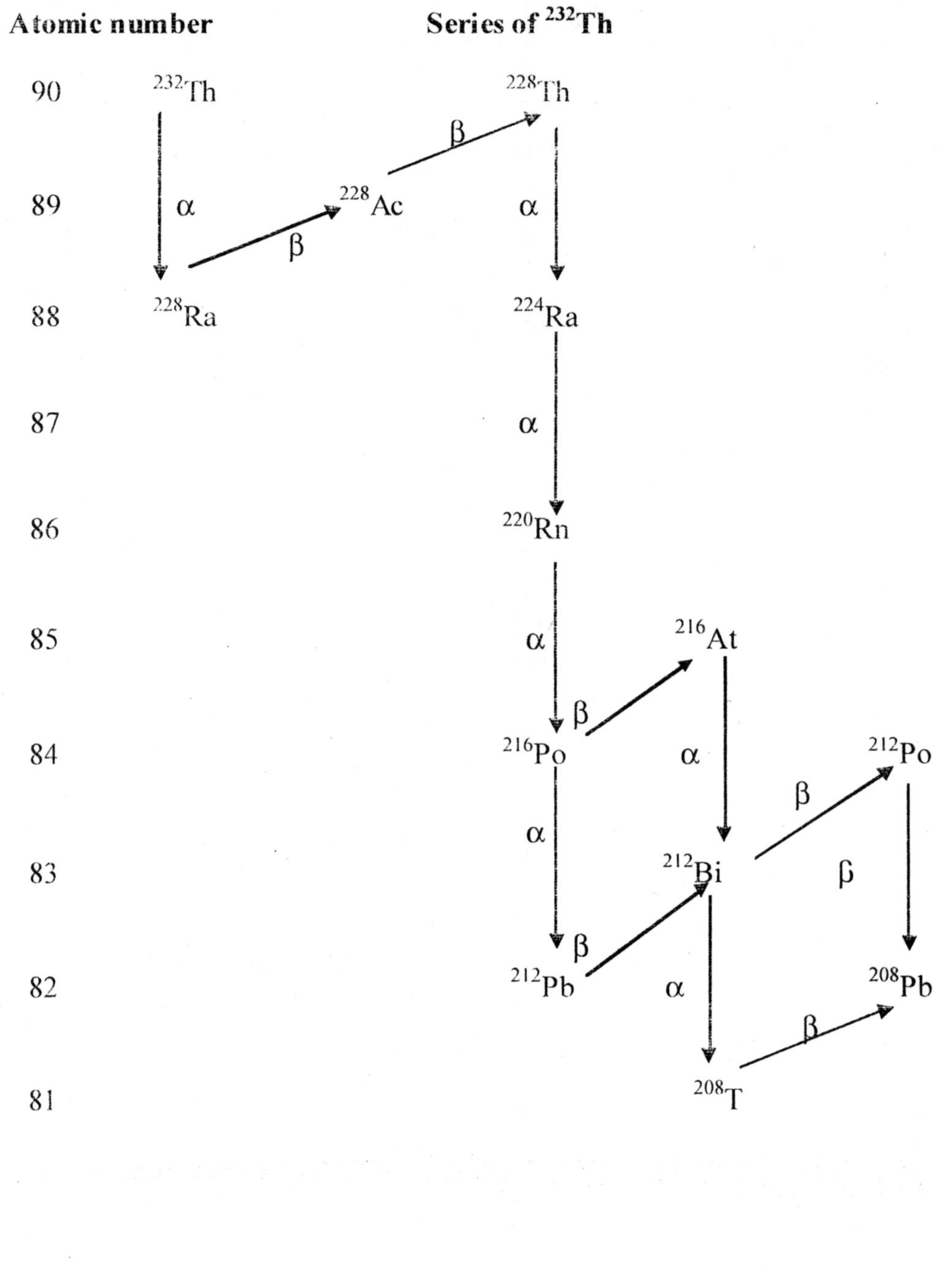

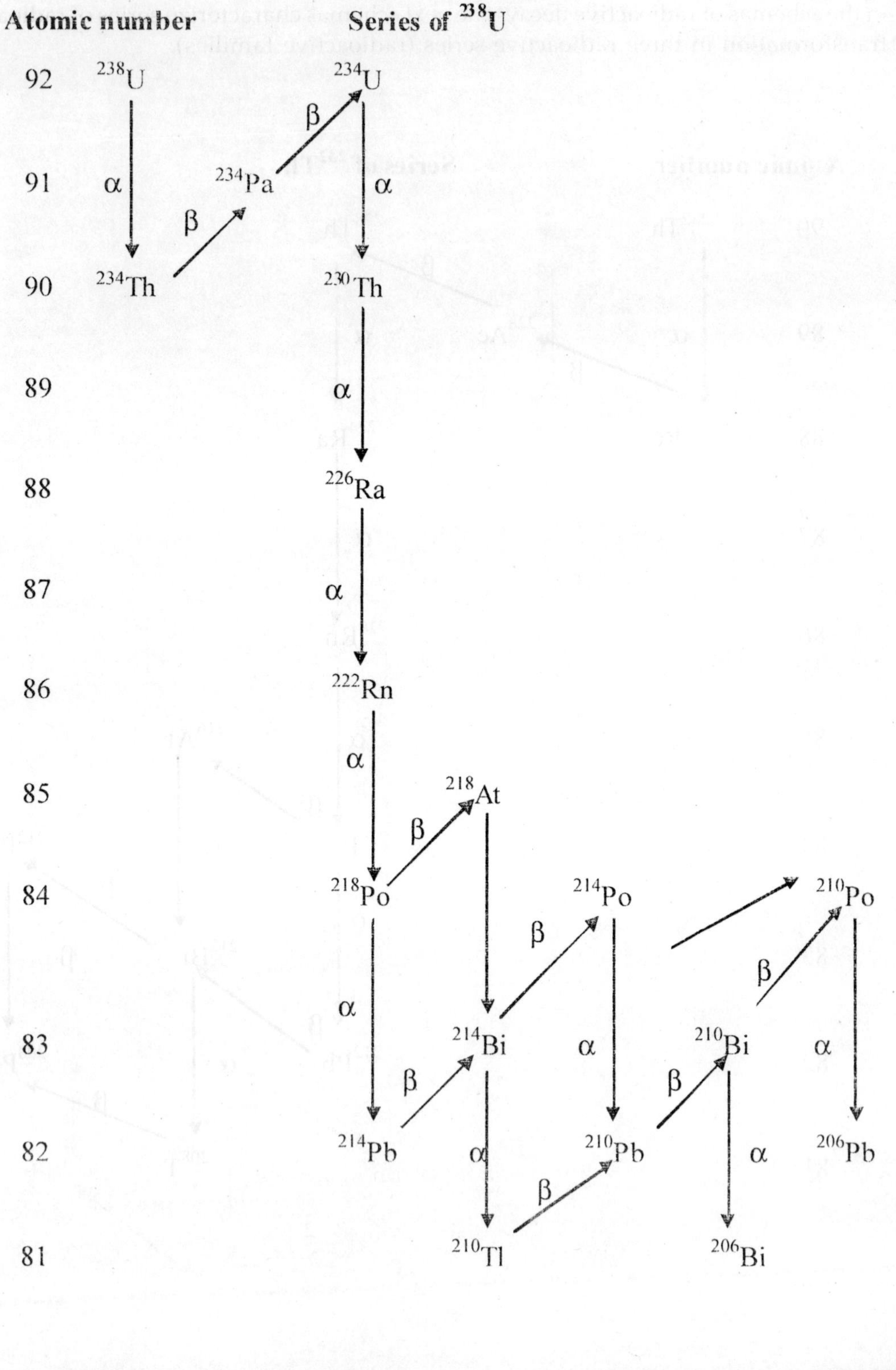
Atomic number
Series of 238U
92
91
90
89
88
87
86
85
84
83
82
81
238U
234Th
234Pa
234U
230Th
226Ra
222Rn
218Po
218At
214Pb
214Bi
214Po
210Tl
210Pb
210Bi
210Po
206Bi
206Pb
α
β

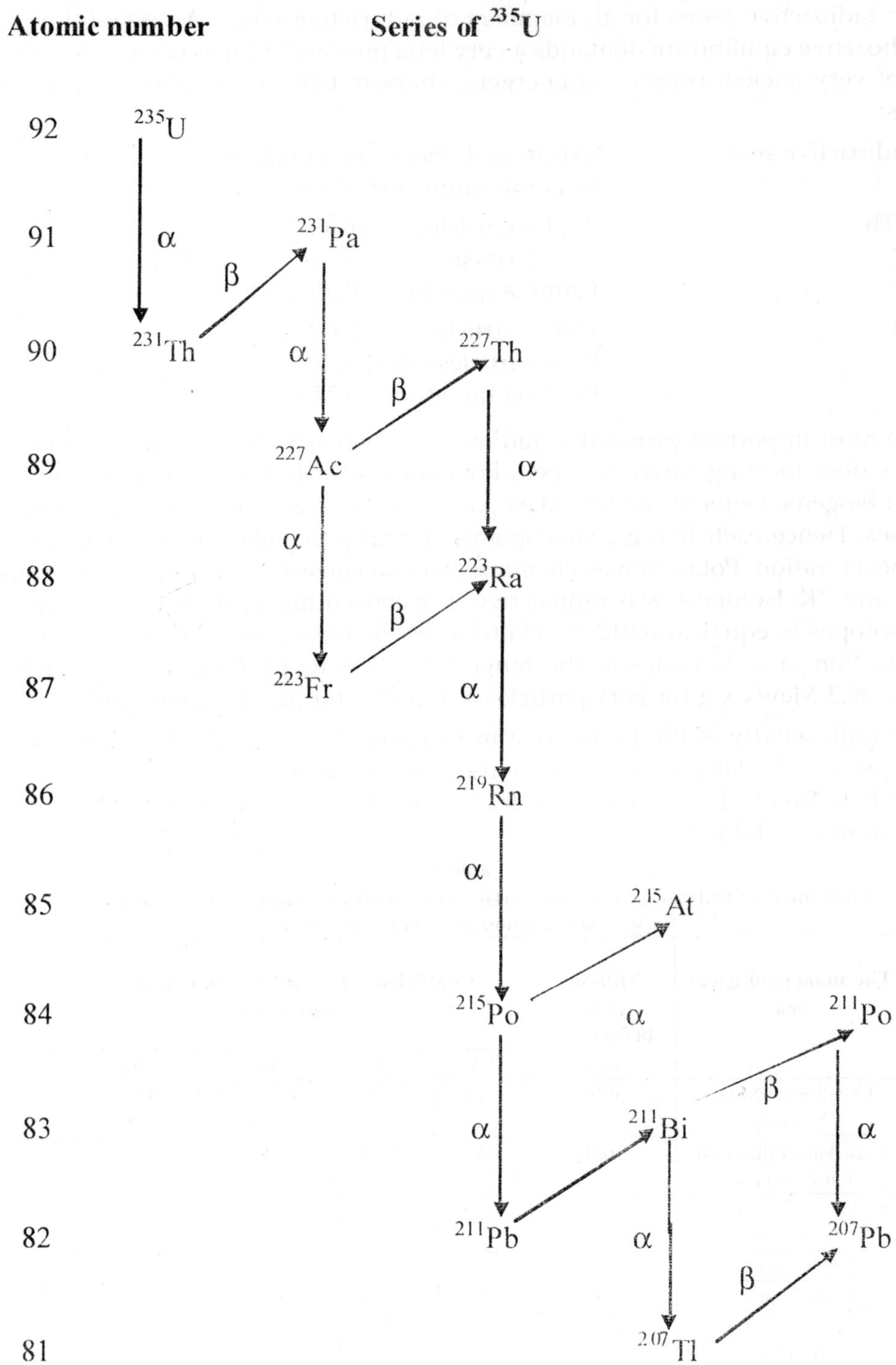
Atomic number
Series of 235U
92
91
90
89
88
87
86
85
84
83
82
81
235U
231Pa
231Th
227Th
227Ac
223Ra
223Fr
219Rn
215At
215Po
211Po
211Bi
211Pb
207Pb
207Tl
α
β

Establishment of equal rate of decay characterizes the radioactive equilibrium state in radioactive series for all members of radioactive series. An establishment of the radioactive equilibrium demands a very long time and it takes place only volcanic rocks of very ancient origin. An energetic characteristic of equilibrium series is as follows:

Radioactive series	Nature and energy of radiation in equilibriums (MeV/s x g)
^{232}Th	Alpha particles - 1.56×10^5 Beta particles - 1.76×10^4 Gamma quantum - 0.93×10^4
^{238}U	Alpha particles - 5.43×10^5 Beta particles - 8.62×10^4 Gamma quantum - 3.27×10^4

The most important primordial radionuclide between the light radionuclides as the main dose-forming native isotope is Potassium-40 - ^{40}K. Potassium is an important mineral biogenic element, which takes part in many biochemical and physiological processes. Hence each living cells contains eternally this element in comparatively high concentration. Potassium as chemical element consists of three isotopes, namely ^{39}K, ^{40}K and ^{41}K. Isotope 40K is radioactive. The concentration of ^{40}K in the mixture of three isotopes is equal to 0,012 %. Half-life of ^{40}K measures 1.28 10^9 years and its concentration in soils varies in the range from 0.37 to 1,1 Bq/g. Energy for ^{40}K is equal to 39.2 Mev/s x g for beta particles and 5.73 – for gamma quantum.

The radioactivity of the Earth crust in Paleozoic Era was essentially higher than now. It can be calculated on the basis of the contemporary radioactivity of Potassium and the half-life of ^{40}K. The curve of decreasing of ^{40}K radioactivity with geological time is shown in Table 5.

Table 5

Abatement of radioactivity in the earth's crust owing to radioactive decay of the long-living radionuclides (Perzov, 1964).

The main geological era	Millions years before our era	Coefficient of arbitrary abatement of radioactivity			
		^{238}U	^{235}U	^{232}Th	^{40}K
Creation of Solar System	5000	2.14	128	1.29	14.3
Creation of the hard earth's crust	2000	1.35	7.05	1.08	2.82
Archaeozoic granites in Finland	1500	1.25	4.3	1.07	2.14
Devonian	400	1.06	1.47	1.025	1.23
Carboniferous era	300	1.05	1.35	1.02	1.17
End of Tertiary period	0.5	1	1	1	1

It is obvious than the level of background was roughly at a tenfold higher than at this geological period. It is not inconceivable that ionizing radiations took part in the physic-chemical processes directly related to biochemical evolution and the first stapes of evolution of species. The formation of some organic compounds from the mixture of carbonic dioxide, ammonia and other simple substances. Mutagenic effect of irradiation could involve in the evolution of species. Other primordial radionuclides are outlined in Table 6.

Table 6
Singly occurring natural radionuclides of terrestrial origin

Radionuclide	Abundance of isotope in element (%)	Half-Life, years	Type and Energy of Radiation (MeV)	Yield of radiation (%)	Specific activity of element (Bq/g)
^{40}K	0.012	1.26×10^{9}	β^- 1.33 γ with EC* 1.46	89 11	31.635
^{50}V	0.25	6×10^{15}	γ with β^- 1.46 γ with EC 1.55	30 70	1.11×10^{-5}
^{87}Rb	27.9	4.8×10^{10}	β^- 0.28	100	8.88×10^{2}
^{115}In	95.8	6.0×10^{14}	β 0.48	100	0.184
^{123}Te	0.87	1.2×10^{13}	EC		0.078
^{138}La	0.089	1.12×10^{11}	β^- 0.21 γ with EC 0.81 1.43	80 70	0.78
^{142}Ce	11.07	$> 5 \times 10^{16}$	α		2.07×10^{-4}
^{144}Nd	23.9	2.4×10^{15}	α 1.38		0.9×10^{-2}
^{147}Sm	15.1	1.05×10^{14}	-		0.13

^{148}Sm	11.27	> 2 x 10^{14}	-		5.01 x 10^{-2}
^{146}Sm	13.82	> 1 x 10^{15}	-		1.2 x 10^{-2}
^{152}Gd	0.20	1.1 x 10^{14}	α 2.1		1.6 x 10^{-3}
^{156}Dy	0.052	> 1 x 10^{15}	-		4.44 x 10^{-8}
^{174}Hf	0.163	2 x 10^{15}	α 2.5		6.2 x 10^{-5}
^{176}Lu	2.6	2.2 x 10^{10}	β^- 0.43 γ 0.089, 0.203, 0.306		88.8
^{180}Ta	0.012	> 1 x 10^{12}	-		0.88 x 10^{-2}
^{187}Re	62.9	4.3 x 10^{10}	β^- 0.003		1.04 x 10^{3}
^{190}Pt	0.013	6.9 x 10^{11}	α 3.18		1.4 x 10^{-2}

*- EC - Electron Capture

The role of these radionuclides as dose-forming factors is exiguous in comparison with ^{40}K and some elements pertaining to the radioactive families of uranium and thorium. Any living beings are necessarily radioactive from the fact that potassium is an obligatory element in whatever biological system.

Cosmogenic Radionuclides

Cosmogenic radionuclides are arising as result of nuclear reaction between atoms of some terrestrial, marine and atmospheric elements and cosmic radiation, which permeate all of space. The source of cosmic radiation is primarily located outside of solar system. This radiation subsists in many forms from high-speed heavy particles as protons, neutrons and nucleus of some elements to high-energy photons and muons. The atoms in the upper atmosphere interact with particles of cosmic radiation and cosmogenic radionuclides are formed as result of nuclear reactions. The cosmogenic radionuclides can have long half-lives but the majority has much shorter half-lives than primordial radionuclides. The main characters of the cosmogenic radionuclides are demonstrated in Table 7.

Table 7
Radionuclides produced by cosmic rays (Eisenbud, 1987)

Radionuclide	Half-life	Troposphere concentration, Bq/ kg air	Radiation and energies in Mev
^{3}H	12 years	0.118	β^- - 0.0186
^{7}Be	53 days	1.04×10^{-2}	K-capture γ 0.477
^{10}Be	1.6×10^{6} years	1.18×10^{-19}	β^- 0.555
^{14}C	5730 years	0.126	β^- 0.156
^{22}Na	2.6 years		β^+ 0.545 γ 1.28
^{24}Na	15.0 hours		β^- 1.4 γ -1.37, 2.75
^{32}P	14 days		β^- 1.71
^{33}P	24 days		β^- 0.246
^{35}S	88 days		β^- 0.167
^{36}Cl	3.1×10^{5} years		β^- 0.714
^{38}S	2.87 hours		β^- 1.1 γ 1.88
^{38}Cl	37 min		β^- 4.91 γ 1.60 2.17
^{39}Cl	55 min		β^- 1.91 γ 0.25 1.27 1.52

THE MOST IMPORTANT NATURAL RADIONUCLIDES

Polonium

Polonium is a radioactive element. It consists of 7 isotopes which arise in ^{238}U, ^{235}U and ^{232}Th series. Among these radioisotopes only ^{210}Po has comparatively prolonged half-life time, namely 138 days. Other radionuclides of this element are short living isotopes. Polonium ^{210}Po as emitter of alpha particles makes up a very radiotoxic factor when its activity attains corresponding values. This radionuclide is found everywhere because it is produced by Radon, which is a very widespread element. Of special note is a high activity of Plutonium in uranium ores, uranium mines and in soils which formed on granites. As radionuclide of high solubility ^{210}Po enters into water, food, living cells and tissue of any organism hence it is revealed in small activities practically in all objects of biota. Plants can accumulate it. Among other plants ^{210}Po is a component of tobacco leaves and cigarette smoke contains this radionuclide in activities, which can produce cancer in human beings and animals. The circulating ^{210}Po cause genetic damages and early death from diseases: liver and bladder cancer stomach ulcer, leukemia, and cardiovascular diseases. It was demonstrated in numerous experiments. It was concluded that cigarette smoking accounts for 30 % of all cancer deaths.

Radon

Radon is a radioactive gas. It is colourless, odourless, tasteless, and chemically inert. Radon consists of three isotopes formed the radioactive decay of ^{238}U, ^{235}U and 232Th. There are twenty known isotopes of radon. The most stable isotope is ^{222}Rn, which is a decay product of ^{226}Ra, has a half-life of 3.823 days and emits alpha particles. ^{220}Rn is a natural decay product of thorium and is called "thoron." It has a half-life of 55.6 seconds and also emits alpha particles. ^{219}Rn is derived from ^{235}U, is called "actinon," is an alpha emitter and has a half-life of 3.96 seconds. As a dose-forming radionuclide ^{222}Rn is the most important among other isotopes of this element. It accompanied uranium and occurs widely in Earth's crust in rock, soil, air and water. Once produced, radon moves through the ground to the air above. Some remains below the surface and dissolves in water that collects and flows under the ground's surface. Radon is a significant contaminant that affects indoor air quality worldwide. Radon gas from natural sources can accumulate in buildings. This gas can be found in some spring waters and hot springs. Radon emanates naturally from the ground all over the world, particularly in regions with soil formed on granite or some shale. Depending on how houses are built and ventilated, radon may accumulate in basements and dwellings.

When radon undergoes radioactive decay, it emits alpha particles and produces short-lived radioactive decay products. The daughter products of decay of Radon are not gases and can easily attach to dust and other particles in air. Those particles can be transported by air and can also be breathed.

The amount of time between exposure and the onset of the disease may be many years.

Breathing radon and products of its decay can induce transformation of cells and cancerogenesis. The study reviewed and evaluated data from many prior studies and drew conclusions support estimates that radon causes about 21,000 lung cancer deaths per year in USA. It is widely agreed that radon exposure is the second leading cause of lung cancer after polonium in cigarette smoke (Eisenbud, 1987). Research suggests that swallowing water with high radon levels may pose risks, too, although risks from drinking water containing radon are much lower than those from breathing air-containing radon. Because most people spend as much as 90 percent of time indoors, indoor exposure to radon is an important concern.

Most indoor radon comes into the building from the soil or rock beneath it.

Radon and other gases rise through the soil and get trapped under the building. The trapped gases build up pressure. Air pressure inside homes is usually lower than the pressure in the soil. Therefore, the higher pressure under the building forces gases though floors and walls and into the building. Most of the gas moves through cracks and other openings. Once inside, the radon can become trapped and concentrated.

^{210}Pb (RaD)

^{210}Pb, with a half-life of 22.8 years, is a member of the ^{238}U series. It arises as a remote product of ^{226}Ra due to the physic-chemical activity of the intermediate gaseous progenitor ^{222}Rn. The gaseous ^{222}Rn escapes into the atmosphere whereupon radioactive equilibrium is sufficiently disturbed. ^{222}Rn, with a half-life of 3.8 days, decays through a series of very short half-life isotopes to lead-210. This process produces excess of ^{210}Pb in the atmosphere and subsequently the hydrosphere. The highly reactive lead is adsorbed on different objects on environment particularly in hydrosphere.

Radiocarbon and Tritium

Carbon-14 (^{14}C) and tritium (^{3}H) are the main dose-forming radionuclides. These radionuclides refer to biogenic elements and amounts of these radionuclides in living beings are much higher than other cosmogenic radionuclides. The radiobiological action of radionuclides involved into biologically important molecules can be related to the transformation of atoms owing to radioactive decay when helium atom arises instead of hydrogen atom or nitrogen atom arises instead of carbon atom apart from direct irradiation.

^{14}C of natural origin is present in the carbon of all living beings at the historically constant specific activity of 0.222 Bq/ g of carbon.

The specific activity of radiocarbon in plant and animal organism can be calculated on the basis of total amount of carbon in the body. The concentration of carbon in organism depends on the water concentration and varies over a wide range from 25 to more than 220 g per kg. The specific radioactivity of ^{14}C in body of organisms may vary within broad limits from 5 to 50 Bq/ kg.

^{40}K

The relationship between the isotopes in whatever polyisotopic elements is constant in all terrestrial and marine objects as well as in meteorites. Therefore radioactive isotope ^{40}K is a necessary accompaniment of potassium. This element is by far the most abundant naturally occurring radioactive element within the cells of any living organisms from microorganisms to human body. Potassium is essential for many biochemical and physiological processes and it can not be substituted by another element. The adult human being contains about 140 g K and daily demand in this element is equal near 2.5 g per day. When ^{40}K decays 89 % of the decay events give rise to the emission of beta radiation with maximum energy of 1.33 Mev and the other 11 % of the decay produce a gamma radiation with energy of 1.46 Mev. The daily serving of potassium for human contains about 80 Bq radioactive ^{40}K. Inasmuch as about 98 % of potassium resides within cells at least 98 % of the atom disintegrations take place inside the cells.

Natural Radioactivity of Environment

Natural radioactive primordial and cosmogenic radionuclides contain in all natural objects from rocks to living beings. These elements appear in air, natural waters, microorganisms, mushrooms, plants and animals. One of the main macroelements without which any alive cell can not exist is potassium having radioactive isotope ^{40}K.

Natural radioactivity is inherent to the earth's crust from the origin of planets. That is why irradiation of all objects was necessary condition in the course of Biosphere evolution. Nevertheless there are no facts supporting an idea that the natural radioactivity of environment is necessary for performing of vital processes though it is known that many factors of chemical and physical nature which are inherent in the earth's crust have come in an arsenal of factors obligatory for alive organisms.

A problem of biological function of a natural radioactivity has not to the present day been decisively solved (Grodzinsky, 1965). However there is a sure basis to count that low doses of irradiation with gamma rays, beta- and alpha- particles should influence as mutagenic factor.

Physicochemical State of the Natural Radionuclides in Environment

Ability for migration in the natural radionuclides depends on their solubility and acidity of soil solution. Potassium exists in soil in fully soluble ionic forms, in the exchangeable forms in soil absorbing complex as well as in a practically non-soluble form as constituents of soil primary minerals. Radiocarbon is found in chalkstones, organic substances, in atmosphere and water in the form of O_2. Tritium is an important constituent of any organic substances and water molecule. Heavy radioactive elements are usually constituents of number minerals. Apart from chemically bound states these radioactive elements can be found in the dispersion states connected with adsorption processes. Ionic forms of the natural radioactive elements are known. These latter can form complex compounds and absorbed forms in soil. Radioactive noble gases – isotopes of radon are constituents of air in the atmosphere, soil and water.

Areas Marked by High Radiation Background

The level of radiation background due to Potassium, Uranium, Thorium, Radium etc is fairly constant over the world being 8 $\times 10^{-8}$ -16$\times 10^{-8}$ Gy/hour. However there are areas in different continents where the level of radiation background in environment is substantially higher in comparison to average value of natural radioactivity. Major anomalies in the natural radioactivity of soils are famed for three countries, Brazil, India and China. The high natural radioactivity in these countries is due to high concentration of radioactive minerals in soils.

In Brazil, the radioactive deposits are of two types: the monazite sand deposits along certain beaches in the states of Esperito Santos and Rio de Janeiro, and the regions of alkaline intrusive in the state of Minas Gerais (Eisenbud, 1987). Monazite is a highly insoluble rare earth mineral. It occurs in some beach sands together with mineral ilmenite, which is characterized by a black color as a consequence these sands have black coloration. The black sands are found along beaches of Atlantic Ocean. The external radiation levels on these beaches range up to 50 μGy/hr. This value of a dose-rate is almost 400 times average background. The major radionuclides in monazite are from ^{232}Th combined with radionuclide of ^{238}U including ^{226}Ra. The high level of radioactivity is observed in Brazilian vacation town Guarapari, which has a stable population about 7000 people. Almost 60,000 inhabitants of the black sands region are exposed to abnormally high radiation dose-rate. Cytogenetic investigations have shown a higher than normal frequency of chromosome aberration in human lymphocytes.

The monazite deposits are more extensive than those in Brazil in the state Kerala, on the Southwest of India, where about 100, 00 persons inhabit this radioactive anomaly. The dose-rate from external irradiation is, on average, 5 - 6 μGy/year though sometimes-individual exposure up to 32.6 μGy may be observed (Eisenbud, 1987). A higher than normal incidence of Down's syndrome has been reported for this region of the high natural radioactivity. There are radioactive anomalies in Iran, where in province Ramsar dose-rates run to 5.5 μGy/hr. Many areas of radioactive anomalies are found in France, Czech Republic, Finland, and Austria. To illustrate, specific radioactivity of water in Bad stain (Austria) reaches 111 Bq/L. The radioactivity of spring water in Helsinki (Finland) due to radon exceeds 629 Bq/L. The dose-rate in air over soil surface in Languedoc (France) is over 0.52×10^{-10} A/kg. A very interesting radioactive anomaly was found in mining region Francevill in Central Gabon (West Africa) near the site of an open-pit uranium mine called Oklo. The share of ^{235}U in the uranium ore in this mine is much less than in normal uranium minerals. This deviation from the normal relation between two main isotopes of uranium can be caused by acting many millions years ago "nature radioactive reactors". ^{235}U was subjected to depletion similar to process, which takes place in artificial nuclear reactor.

Natural Radioactivity of Atmosphere

A number of the natural radioactive elements are as a rule detected in atmospheric air. Five groups of radioactive materials are marked out in air. They are as follows:

1. Radioactive isotopes which are appeared by means of nuclear reaction between atom nucleus and cosmic radiation. Radiocarbon, ^{14}C and tritium, ^{3}H are the main cosmogenic radionuclides in atmosphere.
2. Radioactive gases which are injected in atmosphere through ground surface. Isotopes of radon, namely, ^{222}Ra, ^{220}Rn (toron) and ^{219}Ra (actinone) are the main radioactive gases in air.
3. Air-borne radionuclides adsorbed on dust particles. Dust particles can be formed from uranium and thorium containing rocks. Aerosol particles can also adsorb radioactive atoms, which are products of radioactive decay of radon isotopes.
4. Radioactive materials as a product of condensation processes under influence of cosmic radiation;
5. Radionuclides which appear in air owing to technical activity of people for example due to open winning of uranium.

The content of natural radioactive elements in surface air of atmosphere is very variable parameter which is ranged from 0.0007 to more then 70 Bq/m^3. The highest radon amounts are naturally enough registered in surface air in territories where uranium containing rocks of not too deep bedding are emerged. High radon contents in Ukraine are usually registered in territories located on the Ukrainian Crystalline Shield consisted of granites. In many countries where the underlying geological formation is notable for high uranium content in homes the concentration of ^{222}Rn and radon daughter products is so high that risk for people health can be greater than from many other pollution hazards. A linear relationship between radon exposure and incidence of lung cancer has been observed among uranium miners. Apparently exposure to environmental radon and its daughter products may be responsible for 6 – 12 % of the total number of the lung cancer. Radon can enter the indoor atmosphere in a number of ways mainly by diffusion from building materials and diffusion from soil and ground water through breaches in the foundation. Interest in indoor radon has been developed last decades in many counties.

Natural Radioactivity of Water

Radioactivity of river water depends on chemical composition of the littered rocks and the character of mineralization of water gathered in drain basin. The levels of radioactivity in the main rivers of our planet are displayed in Table 8.

The specific radioactivity in river water varies over a wide range from 18.5 Bq/m^2 in Danube River to 728.9 Bq/m^2 in Rio Grande. The main role in formation of the radioactivity in water takes play, to be sure the radioactivity of bed rocks. The basic part of radioactivity of river water is mainly determined by ^{40}K and Radium. The average amounts of the natural radioactive elements in river water are shown in Table 9.

Table 8

Mineralization and specific radioactivity of water in some rivers (Gorev *et al.*, 1993)

River	Amount of mineral elements, mg/L	Na + K mg/L	Average specific radioactivity, Bq/m3
Amazon (South America)	30.0	3.3	18.5
Parana (South America)	90.6	18.6	107.3
Rio Negro (South America, Argentine)	166.7	13.5	81.4
Colorado (USA)	684.9	102.7	606.8
Mississippi (USA)	210.6	13.8	81,.4
Missouri (USA)	419.8	38.0	225.7
Saint Lawrence (North America)	200.2	7.2	40.7
Rio Grande (USA)	650.9	123.5	728.9
Columbia (North America)	112.0	9.3	51.8
Yukon (Alaska)	135.2	6.0	37.0
Thames (UK)	358.3	12.3	70.3
Main (Germany)	196.1	5.8	33.3
Rhine (Europe)	281.2	5.2	29.6
Elbe (Europe)	188.7	13.7	81.4
Danube (Europe)	331.0	5.3	29.6
Weser (Germany)	301.0	34.2	192.4
Mahanadi (India)	73.3	5.7	33.3
Seraia (Java)	122.3	15.3	92.5
Merano (Java)	116.6	9.4	55.5
Nile (Egypt)	119.1	11.8	66.6

Table 9

Mean concentrations of the natural radioactive elements in river water (Gorev *et al.*, 1993).

Radioactive element	Specific radioactivity, Bq/m^3	Radioactive element	Specific radioactivity, Bq/m^3
K	37	Ra	24.42
U	7.4 - 11.1	Th	8.14

The World Ocean is defined as a tremendous repository of natural radioactive elements. The water of rivers transfers genuinely a great activity of these elements for a long time throughout the whole lifetime of the rivers. Total value of the natural radioactivity accumulated in water is very impressive. The Table 10 gives the content and activity of the main nature radionuclides in seawater.

Table 10

Radioactivity of sea water (Gorev *et al.*, 1993)

Radionuclide	Concentration of radioactive element in water, g/L	Specific radioactivity of radionuclide in water, Bq/m^3	Total radioactivity in the World ocean, $n \times 10^{18}$ Bq
^{40}K	4.5×10^{-5}	12900	16,770
^{87}Rb	8.4×10^{-5}	222	289
^{238}U	2.0×10^{-6}	111	144
^{235}U	1.5×10^{-8}	3	4
^{232}Th	1.0×10^{-8}	0.18	0.23
^{226}Ra	3×10^{-13}	11.1	14.4
^{14}C	4.0×10^{-14}	7.4	9.6
^{3}H	8×10^{-17}	37	48.1

- - A value of World Ocean's volume in 1.3×10^{18} m^3 is used in calculation.

The great bulk of the radioactivity in the World Ocean due to potassium, rubidium and radium. Underground water located in rocks characterized by high radioactivity is exemplified by higher contents of natural radioactive elements. Three general types of radioactive water are correspondingly recognized, namely, radon, radium and uranium water (Eisenbud, 1987). Radon waters are developed in acid igneous rocks. The concentrations of radon in these water exceed 185 Bq/L. Content of Radium is less than $1 \times ^{-10, -11}$ g/L and concentration of Uranium do not exceed 3×10^{-5} g/L. The most widespread radon waters are related to the residual soil and cracks in acid abyssal igneous rocks.

Radium waters are common in sediment rocks where waters are characterized by high mineralization. Content of radium progresses to n × 10^{-11} g/L in carbonate sediment rocks. The concentration of this element reaches (2-12) × 10^{-10} g/L in sand rocks and n × 10^{-9} g/L in metamorphosed shale. Uranium waters are confined to ferriferous shale, sandstone enriched with organic substances and compounds of copper and iron. Waters enriched with high contents of uranium, radium and radon occurs in uranium deposits. These waters are characterized by very high concentration of all natural radioactive elements. The levels of their concentration are as follows: Radon to 185 kBq/L, radium to 9 × 10^{-9} g/L and Uranium to 9 × 10^{-2} g/L. Waters of this type are formed in zones of intensive water cycle in oxidative medium.

With rare exception, content of thorium in natural water is low. The amount of this element in sediment and metamorphosed rocks measures 1 × 10^{-6} g/L. The content of natural radioactive elements reaches very high values in some mineral springs. This is how matters stand in some mineral springs in France where radium content is 1.4 × 10^{-10} g/L, there are mineral springs in Austria (Bad Gastein) where content of radium reaches 1 × 10^{-7} g/L. The content of radium reaches a level of 3 × 10^{-7} g/L in Colorado state (USA) as well as 5 × 10^{-7} in Czechoslovakia (Yahimov uranium rock deposit). The highest amount of radon in waters 66,600 Bq/L on record in Yahimov mineral spring.

Natural Radioactivity in Soils

Activity levels of the natural radioactivity in soil vary greatly depending on soil type, mineral make-up and biogeochemical transformations of radionuclides. Radionuclides contained in the rock matrix are the major sources of the natural radioactivity in soil. Activities of the natural radionuclides in the main rocks, which generate mineral constituents of soil, are shown in Table 11.

Table 11
Average contents of the natural radioactive elements in lithosphere (Gorev *et al.*, 1993).

Element	Contents of elements, %					
	Ultra basic rocks	Basic rocks	Medium rocks	Acid rocks	Sedimentary rocks	Soils
K	5.10^{-1}	$8.3.10^{-1}$	2.31	3.34	2.28	1.36
Rn	$6.5.10^{-18}$	$1.7.10^{-16}$	$3.9.10^{-16}$	$7.6.10^{-16}$	$6.9. 10^{-16}$	
Ra	1.10^{-12}	$2.7.10^{-11}$	$6. 10^{-11}$	$1.2.10^{-10}$	$1. 10^{-10}$	$8. 10^{-11}$
U^{235}	2.10^{-8}	6.10^{-7}	$1.3.10^{-6}$	$2.5. 10^{-6}$	$2.3. 10^{-6}$	
Th	6.10^{-4}	$3. 10^{-4}$	$7. 10^{-4}$	$1.8.10^{-3}$	$1.1. 10^{-3}$	$6. 10^{-4}$
U	3.10^{-6}	8.10^{-5}	$1.8. 10^{-4}$	$3.5. 10^{-4}$	$3.2. 10^{-4}$	$1. 10^{-4}$

Uranium, thorium, members of their radioactive families and potassium are components of the natural radioactivity in soil. The radioactive equilibrium is disturbed when radionuclides are migrating from the particles of maternal rocks into soil solution or the secondary minerals.

The levels of the natural radioactivity in different types of soil are shown in Table 12.

Table 12

Average content of the major natural radionuclides in soils (Moiseev and Ivanov, 1990)

Soil	Average specific activity, Bq/kg			Dose rate in air, nGy/hr
	^{40}K	^{238}U	^{232}Th	
Sierozems	670	31	48	74
Tape soils	700	28	41	69
Chestnut soils	550	27	37	60
Chernozems	410	22	36	51
Grey forest soils	370	18	27	41
Sod-podzol soils	300	15	22	34
Podzol soils	150	9	12	18
Peaty soils	90	6	6	11
Average value for world	370	26	26	46
Characteristic range	110 - 740	11 - 52	7.5 - 48	14 - 90

An attention should be given to the leadership of potassium in the radioactivity of without exception all types of soil. Thorium ranks the second place in the radioactivity of soil. Since radium is an obligatory radioactive component in soil radon is permanently emanated from soil into atmosphere. Owing to the radioactivity of soil corresponding levels of dose rates of the ionizing radiation are permanently sustained in air over soil surface. The dose-rate can be calculated on the basis of the specific radioactivity of soil. The consequent data are shown in Table 13.

Table 13

Dose-rate in air on the height 1 m above soil surface related to potassium, uranium and thorium

Radinuclide	Dose-rate on unit of specific radioactivity in soil, 10^{-10} Gy/hr
^{40}K	0.43
^{238}U	4.27
^{232}Th	6.62

The total activity of the natural radionuclides in plough-layer is very impressive. It is seen from the calculated data what are listed in Table 14.

Table 14

Average content of nature radioactive elements in soil of Ukraine (Grodzinsky, 1965)

Soil	Collecting ground	Content		
		Th, $n \times 10^{-4}$ %	U, $n \times 10^{-5}$%	Ra, $n \times 10^{-10}$%
Sod podzol	Zhitomir region	12.3	14.7	-
Gray forest podzolized	Vinnitza region	9.3	7.7	0.5
Chernozem podzolized	Charkov region	12.0	24.5	1.0
Chernozem ordinaries	Kirovograd region	7.3	7.7	1.2
	Dnepropetrovsk region	13.7	13.3	0.8
	Donetzk region	16.4	25.2	1.5
	Lugansk region	7.3	16.1	0.9
Chernozem vigorous	Kiev region	10.3	16.1	2.1
	Sumy region	14.7	13.3	0.9
	Cherkassy region	11.0	12.6	1.1
	Ternopol region	12.7	15.4	1.2
Chernozem southern	Zaporozhye region	8.1	16.1	0.9
Chernozem leached	Donetzk region	8.2	23.8	0.8

Chernozems ordinary and vigorous are the most radioactive soil. The high amount of organic matter, which is capable of accumulation of radioactive substances, is the possible cause of the augmentation of the natural radioactivity characteristic of this type of soil.

Specific beta-radioactivity of soil reflects the total amount radionuclides in this natural object. The highest level of beta radioactivity is typical of chernozems and least level is characteristic of peat bog where concentration of potassium is commonly very low. The Table 15 lists the value of the beta-radioactivity in plough layers of some soils.

There are two provinces in Ukraine with soils of high radioactivity where the underlying bedrocks are presented by granites characterized by high contents of uranium and radium. The highest level of natural radioactivity in soil was located in granite province where the radioactivity of soil in transitional horizon was 947 B/kg. Level of radioactivity is usually low for swamplands.

Alpha-radioactivity of soils averages between 0.6 and 5.25 Bq/g. It is basically checked with beta-radioactivity of soil.

Data on the concentration of uranium in soil support a thought on the high radioactivity inherent to chernozems. The data on the content of uranium in soil are demonstrated in Table 16.

Table 15

Total beta- radioactivity of plough layers of some soils in Ukraine (Grodzinsky, 1965).

Soil	Beta radioactivity, Bq/kg
Low-lying peat bog	37 - 185
Sod podzol	140
Gray forest podzolized	92
Gray forest	77 – 388
Sod moderately podzolized	381
Dark gray forest	617
Meadow chernozem	718
Chernozem vigorous carbonate	703 – 947
Chernozem vigorous leached	773 – 902

Table 16

Content of Uranium in plough layer of some soils in Ukraine (Grodzinsky, 1965)

Soil	Content of uranium, %
Solonetz alkalinized heavy-loamy	2.5×10^{-5}
Chernozem south medium-loamy	2.5×10^{-5}
Chernozem south heavy-loamy	3.5×10^{-5}
Chernozem ordinaries medium-humic	3.5×10^{-5}
Chernozem vigorous slightly-humic	4.0×10^{-5}
Chernozem ordinaries eroded	4.0×10^{-5}
Dark-gray podzolised	2.5×10^{-5}

In the case of deterioration in the radioactive equilibrium in soil the relationship between radium and uranium should be varied. The values of this relationship in soil are depicted in Table 17.

The features of distribution of radioactivity along the horizons of soil depend on type of soil and character of underlying bedrocks. Though radioactivity of soil is commonly constant along the profile.

There is a correlation between content of humus and uranium:

Content of humus, %	Content of Uranium, %
Cherokee vigorous	
5.67	4.0×10^{-5}
6.21	5.0×10^{-5}
Chernozem ordinaries	
3.21	9.1×10^{-5}
3.04	12×10^{-5}

Table 17

State of radioactive equilibrium in some soils (Grodzinsky, 1965)

Type of soil	Genetic Horizon	Depth of sampling, cm	Th/U	Ra/U × 10^{-7}
South chernozem	$A_{tillable}$	0 – 29	5.0	6.0
	A	29 – 52	5.2	5.6
	B	52 – 90	4.5	5.1
Chernozem eroded	$A_{tillable}$	0 – 30	3.4	3.6
	C	30 – 50	3.8	2.5
Chernozem ordinaries	A	0 – 39	4.5	6.0
	B	39 – 62	5.0	6.4
	C	62 – 89	2.5	1.8
Chernozem vigorous low-humus	$A_{tillable}$	0 – 30	6.3	12.8
	A	30 – 50	5.0	6.3
	AB	50 – 77	6.4	6.3
	B	77 – 96	6.3	8.0
Chernozem polsolized	$A_{tillable}$	0 – 30	4.8	4.5
	A	30 – 56	8.4	6.0
	B	56 – 80	9.2	6.4
	C	80 – 95	9.5	8.0

The highest concentrations of Uranium located in soil formed on granites or some carbonaceous shale. The contents of Thorium in soils of Ukraine average between 7.3×10^{-4} and 16.4×10^{-4} %.

Fertilizers as Sources of Natural Radioactivity in Soil

It is common knowledge that many fertilizers contain elevated levels of the natural radioactive elements among them potassium, uranium and thorium series. Potassium fertilizers contain isotope ^{40}K and ^{87}Rb because potassium and rubidium have similar geological origin in rocks.

The radioactivity of some potassium fertilizers is as follows:

Potassium sulphate - 12,230 kBq/kg
Potassium chloride - 10,210 kBq/kg
Silvinite - 7,416 kBq/kg

Using of potassium fertilizers gives not too much in the total radioactivity of soil which contains potassium in high concentration: an augmentation of radioactivity due to potassium fertilizers should not be exceeded 0.04 % in chernozems and no more than 4 % in peat soils.

The phosphorous fertilizers are produced on a basis of the phosphate rock. The latter can contain elements of uranium series and to a lesser degree thorium series. Uranium and phosphorites are characterized by similar paragenesis and in consequence phosphate fertilizers, which are produced on the basis of phosphorite, contain necessarily elements of uranium series.

The measurement provided in NCRP indicates that the raw phosphate material (super phosphate, diammonium phosphate, phosphoric acid and other.) incorporated into fertilizers contain:

^{238}U - 800 - 5,200 Bq/kg
^{230}Th - 200 - 16,000 Bq/kg
^{226}Ra - 25 - 900 Bq/kg
^{232}Th - 5 - 170 Bq/kg

Many millions tons of agricultural fertilizers are used in all countries and high total radioactivity is introduced in soil with phosphate fertilizers.

There are two pathways by which fertilizers can contribute to the radiation exposure of the people: the external exposure due to gamma radiation emitted by the fertilizers and the internal exposure resulted from the consumption of food manufactured from crops cultured on land that had been fertilized. NCRP Report 95 estimated that the resulting exposure to an average human in the USA might be on the order of 0.01 to 0.02 mSv per year.

Natural Radioactivity in Plants

The natural radioactive elements are found in any plant tissue. Plants from soil, bedrocks and air metabolically absorb these elements. It is significant that the natural radioactivity makes an integral part of the environment from the incipience of the Life on the Earth. A famous Russ9in scientist of the previous century Vladimir Vernadsky wrote: "The Life in the Biosphere emanates from two sources of energy- the sun light and atomic radioactive energy". True, the level of the radioactive energy is on many magnitudes less than the energy of the sun light. Marie Curie has noted that plants are normally are inhabited in radioactive surroundings and this fact can be considered as a motive to believe that the total radioactivity though very small is one from the range of condition which is necessary for the normal development of plants.

An accumulation of radioactivity by plants has been studied for many years from the beginning of the last century. Data obtained in many laboratories show that plants have the property of absorbing various radioactive elements.

We investigated the natural radioactivity in more than 300 plant species related to different phylogenetic groups. These investigations were carried out in years when the level of radioactive contamination with artificial radionuclide was very small.

The value of the total alpha-radioactivity in these plants is demonstrated in Table 18.

The data derived in this Table indicate that the natural radioactivity of plants is beyond all doubt related to the phylogenetic position of taxon. The higher plants are characterized by comparatively low radioactivity. The amounts of radium in plant tissue are as a rule much less than in soils or bedrocks. The data on the content of radium in plants are given in Table 19.

The concentration of radium in plants appears to be proportional to the amount of calcium. Much higher activity of radium should be forthcoming in calciphilous

Table 18

Total alpha radioactivity of some plants (Grodzinsky, 1965)

Plant species	Collecting ground	Alpha-radioactivity in ash, Bq/kg
	Angiosperm	
Vetch (Vicia lutea L.)	The Crimea	0.22
Common maple (Acer campestre L.)	The Crimea	0.92
Olive (Olea europaea L.)	The Crimea	1.59
Durmast oak (Quercus petraea Liebl.)	The Crimea	1.59
Evergreen magnolia (Magnolia grandiflora L.)	The Crimea	1.70
Bilberry (Vaccinium myrtillus L.)	Transcarpathian region	1.92
Holm oak (Quercus ilex L.)	The Crimea	2.02
Hawthorn (Crataegus taurica Pojark.)	The Crimea	2.02
Girlish daisy (Pyrethrum parthenium Smith.)	The Crimea	2.40
Cornelian cherry (Cornus mas L.)	The Crimea	2.59
Strawberry tree (Arbutus andrachne L.)	The Crimea	3.95
Greenweed (Genista albida Willd.)	The Crimea	4.03 – 8.36
Sorrel (Rumex multifidus L.)	The Crimea	4.40
Rock jasmine (Androsace taurica Ovcz.)	The Crimea	4.48
Thyme (Thymus lovyanus Klik.)	Polessye of Ukraine	5.66

Gymnosperms		
Mammoth tree (Sequoiadendron giganteum Lindl.)	The Crimea	0.92
Abies (Abies numidica Lannoy)	The Crimea	1.74
Juniper (Juniperus excelsa M.B.)	The Crimea	1.89 – 3.48
Italian cypress (Cupressus sempervirens L.)		2.04
White cedar (Chamaecyparis lawsoniana Parl.)	The Crimea	
Yew (Taxus baccata L.)	The Crimea	2.29
Stone pine (Pinus pinea L.)	The Crimea	2.41
Gerard pine (Pinus gerardiana Wall.)	The Crimea	2.70
Crimean pine (Pinus pallasiana Lamb.)	The Crimea	2.78
Scotch pine (Pinus silvestris L.)	The Crimea	4.70 – 6.92
Pine (Pinus mughus Scop.)	Transcarpathian region	10.03
Pteridophyta		
Rock brake (Polypodium vulgare L.)	Transcarpathian region	11.36
Bryophyta		
Hypnum cupressifdorme Hedw.	Transcarpathian region	12.03 – 12.51
Rhytidiadelphus triquetrus (Hedw.) Warnst.	The Crimea	12.99
Hylocomium splendens (Hedw.) Br. Eur.	The Crimea	13.28
Tortella tortuosa (Turn.) Gimpr.	The Crimea	14.65
Isothecium myurum (Poll.) Brid.	The Crimea	18.61
Antitrichia curtipendula (Hedw.) Brid.	The Crimea	20.20
Ctenidium molluscum (Hedw.) Brid.	The Crimea	26.94
Hypnum cupressiforme Hedw.	The Crimea	32.71
Lichen		
Dermatocarpon miniatum (L.) Mann	The Crimea	9.91
Peltigera rufescens (Weis.) Humb.	The Crimea	14.31
Cladonia convolute (Lam.) Suza	The Crimea	15.87
Parmelia acetabulum (Neck.) Duby	The Crimea	16.44
Ramalina farinacea (L.) Ach.	The Crimea	17.13
Cetraria glauca (L.) Ach. F. coralloidea (Wallr.) Koerb.	The Crimea	18.75
Anaptichia ciliaris (L.) Koerb		
Ramalina strepsilis (Ach.) A.Z	The Crimea	18.87
Evernia prunastri (L.) Ach.	The Crimea	19.65 – 36.63
Ramalina fraxinea (L.) Ach.	The Crimea	25.53
Evernia furfuracea (L.) Mann.	The Crimea	25.53
Lobaria verrucosa (Huds.) Hoffm.	The Crimea	28.08
Usnea dasypoga (Ach.) Nyl.	The Crimea	28.32
Gyrophora cylindrical (L.) Ach.	The Crimea	44.03
	Transcarpathian region	86.95

Table 19

Concentration of radium in dry mass of plants referring to different phylogenetic groups (Grodzinsky, 1965).

Plant species	Concentration of Radium, Bq/kg	Plant species	Radioactivity of Radium, Bq/kg
		Dicotyledons	
Touch-me not (Impatiens sp.)	0.33	Forget-me-not (Myosotis palustris)	1.29
Crepis (Crepis paludosa)	0.44	Geranium (Geranium palustre)	1.29
Greater celandine (Chelidonium majus)	0.59	Berteroa (Berteroa incana)	1.33
Ox-eye daisy (Matricaria inodora)	0.70	Buttercup (Ranunculus acris)	1.52
Spike loosestrife (Lythrum salicaria)	0.74	Wild madder (Galium mallugo)	1.59
Heal-all (Prunella vulgaris)	1.07	Golden loosestrife (Lysimachia vulgaris)	1.62
Wild chervil (Antriscus sylvestris)	1.29	Common Cow wheat (Melampyrum pratense)	1.55
		Monocotyledons	
Soldier (Stratiotes sp.)	0.11	(Sparganium sp.)	0.26
Broad-leaved cat's tall (Thypha latifolia)	0.15	Water plantain (Alisma plantago)	0.37
Old-world arrowhead (Sagittaria sagittifolia)	0.22	Wild calla (Calla palustris)	0.48
Bermuda grass (Cynodon dactylon)	0.26	Bulrush (Scirpus lacustris)	0.59
Rush flower (Butomus umbellatus)	0.30	Angle onion (Allium angulosum)	0.85
Common reed grass (Phragmites communis)	0.33	Unbranched bur-reed (Sparganium simplex)	1.04
		Gymnosperms, ferns and fungi	
Low juniper (Juniperus sibirica)	0.30	Cepe (Boletus edulis)	0.08
Ostrich fern (Matteuccia struthiopteris)	1.00	The contents of radium to be apparently proportional to contain of calcium in plants. Some higher activity of radium in plants should be forthcoming in calciumphylTinder fungus (Fomes fomentarius)	0.62
		Algae	
Sea wand (Laminaria sp.)	0.11	Rockweed (Fucus sp.)	0.44
Rockweep (Ascophyllum sp.)	0.24	Kelp (Macrocxystis sp.)	1.18
Wing kelp (Alaria sp.)	0.41		

species. The biological influence of uranium can be tied not only to its radioactivity but to its chemical properties.

The amounts of uranium in plants are demonstrated in Table 20.

Table 20

The content of uranium in plants (Grodzinsky, 1965)

Plant species	Content of uranium in ash, n x10^{-7} g/g
Angiosperms	
Crimean hawthorn (Crataegus taurica Pojark.)	0.10
Cornelian cherries (Cornus mas L.)	0.28
Olive (Olea europaea L.)	0.62
Crimean goldendropp (Onosma tauricum Pall.)	0.74
Evergreen magniolia (Magnolia grandiflora L.)	1.02
Cherry laurel (Laurocerasus officinalis L.)	3.00
Red strawberry madrona (Arbutus andrachne L.)	5.84
Holm oak (Quercus ilex L.)	6.72
European beech (Fagus silvatica L.)	10.10
Gymnosperms	
Yellow pine (Pinus ponderosa Dougl.)	0.36
Colorado spruce (Picea pungens Engelm.)	0.95
Mexican pine (Pinus montezumae Lamb.)	1.06
Crimean pine (Pinus pallasiana Lamb.)	1.12
Redwood bur (Sequoia sempervirens Engl.)	3.30
Stankevitch pine (Pinus stankeviczi Fom.)	3.30
Scotch pine (Pinus silvestris L.)	3.85
Gerarda pine (Pinus gerardiana Lamb.)	4.85
Mammoth tree (Sequoiadendron giganteum Lindl.)	4.85
Hymalayan cedar (Cedrus deodara Loud.)	5.00
White cedar (Chamaecyparis lawsoniana Parl.)	13.60
Pteridophytes	
Rock brake (Polypodium vulgare L.)	2.00
Bryophyta	
Hylocomium splendens Hedw.	11.20
Rhitidiadelphus triquetrus (Hedw.) Warnst.	2.00
Sphagnum acutifolium Ehrh.	7.00
Isothecium myurum (Poll.) Brid.	10.00
Tortella tortuosa (Turn.) Gimpr.	11,10
Antitrichia curtipendula (Hedw.) Mitt.	27.50
Ctenidiummolluscum (Hedw.) Brid.	35.00
Hypnum cupressiforme (Hedw.). Mitt (from the diffuse halation of uranium orebody)	44000
Lichen	
Evernia prunastri (L.) Ach.	0.68
Usnea dasypoga (Ach.) Nyl.	0.74
Ramalina farinacea (L.) Ach	0.93
Cladonia convolute (Lam.) Suza	2.50
Ramalina strepsilis (Ach.) A.Z.	2.50
Peltigera rufescens (Weis.) Humb. F. praetextata (Floerk.) Oxn.	3.00
Lobaria verrucosa (Huds.) Hoffm.	3.50
Ramalina fraxinea (L.) Ach.	3.50
Evernia furfuracea (L.) Ach.	4.00

As may be seen from these data some higher content of uranium is typical to mosses. It is most pronounced at the case of moss inhabited on substrate characterized by a high level of uranium.

The total beta radioactivity of plants accounts for the accumulation of the different beta emitting radionuclides in plant tissue. The data on the beta radioactivity in ashes of plants are listed in Tables 21 and 22.

Table 21

Specific beta- radioactivity of ashes in some moss species (Grodzinsky, 1965)

Moss species	Collecting ground	Specific activity, kBq/kg
Polytrichum formosum Hedw.	Transcarpathian region	11.34
Hypnum cupressiforme Hedw.	Transcarpathian region	17.32 – 72.89
Camptothecium lutescens (Hedw.) Breur	The Crimea	20.98
Tortella tortuosa(Turn.) Gimper	The Crimea	23.74
Isothecium myurum(Poll.) Brid.	The Crimea	24.12
Hylocomium splendens (Hedw.) Breur.	Transcarpathian region	27.79 – 106.93
Rhytidiadelphus triquetrus (Hedw.) Warnst.	The Crimea	28.12
Antitrichia curtipendula (Hedw.) Brid	The Crimea	31.01
Hylocomium splendens (Hedw.) Breur	The Crimea	33.00
Ctenidium molluscum (Hedw.) Mitt.	The Crimea	36.48
Sphagnum acutifolium Ehrh.	Transcarpathian region	42.18 – 100.64
Hypnum cupressiforme Hedw.	The Crimea	53.65

The total radioactivity of mosses and lichen is much higher relative to radioactivity of much younger taxons of angiosperm species. It is apparent that an accumulation of elevated amounts of beta emitting radionuclides reflects some specific demands of the ancient plant species in these natural radionuclides.

We can safely assume that an accumulation of radionuclides by plants is distinguished by a specific selectivity. This selectivity is well seen in distinctions between radioactivity in wood tissue and lichen growing on tree branches. These distinctions are shown in Table 23.

Though ^{40}K is apt to be basic beta emitting natural radionuclide a share of this isotope in the total radioactivity of plants may not be a deciding radioactive factor in some old taxones.

Table 22

Specific beta radioactivity of ash in some lichen species (Grodzinsky, 1965)

Lichen Species	Collecting Ground	Specific Radioactivity, kBq/kg
Anaptychia ciliaris (L.) Koerb.	The Crimea	25.53 – 52.54
Bryopogon implexus (Hoffm.) Elenk.	The Crimea	172.42
Cetraria glauca (L.) f. coralloidea (Wallr.) Koerb.	The Crimea	43.18 – 53.21
Cetraria islandica (L.) f. maculata (Vain.) Sav.	The Crimea	33.67 – 78.44
Cetraria islandica (L.) f. maculate (Vain.) Sav.	Transcarpathian region	150.22
Cladonia convolute (Lam.) Suza	The Crimea	19.46
Cladonia pixidata (L.) Fr.	The Crimea	48.66
Cladonia rangiferina (L.) Webb.	Transcarpatian region	49.40 – 80.03
Cladonia rangiformis Hoffm. Var. pungens (Ach.) Vain	The Crimea	48.47
Dactylina madreporiformis (Ach.) Tuck.	The Crimea	26.49
Evernia furfuracea (L.) Mann.	The Crimea	48.84 – 87.69
Evernia furfuracea (L.) Mann.	Transcarpathian region	80.29
Evernia prunastri (L.) Ach.	The Crimea	33.70 – 64.31
Gyrophora cylindrical (L.) Ach.	Transcarpathian region	110.26
Lobaria verrucosa (Huds.) Hoffm.	The Crimea	33.74 – 48.95
Parmelia acetabulum (Neck.) Duby	The Crimea	17.83 – 23.12
Peltigera rufescens (Weis.) Humb.	The Crimea	70.12
Ramalina farinacea (L.) Ach.	The Crimea	56.80 – 78.44
Ramalina strepsilis (Ach.) A.Z..	The Crimea	20.76 – 27.57
Usnea dasypoga (Ach.) Nyl.	The Crimea	36.19 – 88.43

Part of potassium as radioactive element in the total natural radioactivity of plants is commonly lesser than in more recent higher plants. It can be seen in data displayed in Table 24.

The total radioactivity of plants depends on the radioactivity of bed rocks. It can be seen in data derived in Table 25. The radioactivity of plants growing on limestone is very variable.

Possible Role of Natural Radioactive Background in Processes in Biological Systems

An idea on necessity of the natural radioactive background of ionizing radiation in vital activity has a long history from the beginning of the former century. But numerous experimental attempts to find out an impact of the natural radioactive background on biological objects encounter the large difficulties. Really, making an

Table 23

Radioactivity of some tree species and lichen growing on their branches (Grodzinsky, 1965)

Lichen species	Tree species	Beta-radioactivity of ash Bq/g	
		Lichen	Wood
Cetraria glauca (L.) Ach. f. coralloidea (Wallr). Coerb.	European hornbeam (*Carpinus betulus* L.)	45.7	18.2
- " - "	Scotch pine (*Pinus silvestris* L.)	45.7	26.3
Evernia furfuracea (L.) Mann	Crimean pine (*Pinus pallasiana* Lamb.)	48.8	27.2
- " – " – "	Scotch pine (*Pinus silvestris* L.}	49.9	27.7
Anaptychia ciliaris (L.) Koerb.	*Cornelian cherries* (Cornus mas L.)	52.5	10.5
Evernia prunastri (L.) Ach.	Oleaster pear (*Pyrus elaeagnifolia* Pall.)	64.2	21.8

appropriate control is very hard-hitting affair. This is the case: experiments on the finding out of a role of the natural radioactive background in vital activity should be based on two approaches. On the one hand experiments should be carried out in the deep excavation of ore with very scanty radioactivity to reduction of effect of cosmic rays. On the other hand, feeding mineral food staff should contain potassium without ^{40}K. However, it is not possible to change natural potassium on other no radioactive element well as to get potassium without radioactive isotope ^{40}K and to isolate of experimental organisms completely from the action of cosmic rays. In consequences of these difficulties very few experiments have been carried out on this subject.

Any effects were not observed in experiments carried out by Vinogradov with a replacement of potassium of usual natural isotopic composition on potassium without ^{40}K. A fungus *Aspergillus niger* was used in these experiment. Nevertheless existence an influence of decreasing intensity of the radioactive background has been quasi demonstrated in *Protozoa - Paramecium caudatum* and *Paramecium aurella.* Chambers with "low-background (chambers with lead walls of 10 centimeter thickness) have been based on depth 200 m in salt rocky mine. Data received in this experiment showed negative impacts of decreasing background on vitality of paramecium. But nevertheless it was far up to full inactivation of cell population in this experiment. Obviously, the problem of the role of the natural radioactive background in vital activity demands the further researches.

Table 24

Share of the potassium radioactivity in total radioactivity of some plant species (Grodzinsky, 1965)

Plant species	Content of potassium in ash, %	Beta-radioactivity of ash, Bq/g		Share of ^{40}K in total beta-radioactivity, %
		Total	From ^{40}K	
The higher plants				
Vetch (Vicia tenuifolia Roth.)	32.3	14.39	8.66	60.0
Star-of-Bethlehem (Ornithogalum pyrenaicum L.)	22.6	11.40	6.22	54.5
Clear-eye (Salvia sclarea L.)	21.2	10.47	5.92	56.5
Prunella (Prunella laciniata L.)	19.5	14.47	5.48	38.0
Scrophula plant (Scrophularia canina L.)	19.5	17.94	5.48	30.5
Pine stankevicha (Pinus stankeviczi Fom.)	13.0	26.79	3.51	13.1
Mosses				
Tortella tortuosa (Turn.) Gimpr.	3.59	23.75	1.14	4.8
Hylocomium splendens (Hedw.) Br. Eur.	3.16	32.93	1.00	3.0
Hypnum cupressiforme Hedw.	3.40	53.65	1.11	2.1
Lichen				
Cladonia convoluta (Lam.) Suza	1.54	19.76	0.48	2.4
Cetraria islandica (L.) Ach. f. maculate (Vain.) Sav.	1.21	35.89	0.41	1.1
Anaptychia ciliaris (L.) Koerb.	2.29	52.54	0.63	1.2
Dactylina madreporiformis (Ach.) Tuck.	1.01	54.39	0.33	0.6

Russian radiobiologist A. Kuzin stated that the natural radioactive background is an inalienable part of environment which is necessary for the living organisms. A. Kuzin surmised a mechanism of biological action of low doses of ionizing radiation: absorption of quantum by biopolymer molecule within cell leads to forming a long existing exciton named polaryton. Degradation of polaritones is accompanied with an origin of low intensity lighting. This light radiation reminds the mitogenic rays identified by A.G. Gurvitch in twenties of the last century. According to A.G. Gurvitch these rays stimulate cell divisions and growth of cell population in prokaryotic and eukaryotic organisms. Radiation originated from disintegration of polaritones nascent under influence of the natural radioactive background has been named by A. Kuzin as the secondary biogenic radiation.

Table 25

Beta radioactivity of greenweed (*Genista albida* Willd.) growing on different bed rocks in the Crimea (Grodzinsky, 1965)

Bed rock	Radioactivity of ash, Bq/g
Shale	25.05
Limestone	27.6 – 58.09
Liparite	27.6
Volcanic breccias	27.82
Oxykeratophyre	27.94
Diorite	44.77

An energy absorbed by cells from the natural radioactive background is very low therefore it is unlikely that presumable effects of the radioactive background on the vital activity have an energetic nature it can rather be mediated through information processes.

Natural Radioactive Background and Evolution of Species

The level of the natural radioactivity in the time of biochemical off-cell evolution was noticeable higher than in this spell and an ionizing radiation was no doubt a powerful factor taking part in early steps of origin of organic substances. Synthesis of low molecular organic compounds in gases of the primary atmosphere, polymerization of monomers leading to formation of protein-liked molecules and other radiation chemistry processes formed a basis of a pre-cellular evolution of a life on the Earth. The radiochemical reactions of this kind have been reproduced in model systems.

Later presence of an eternal field of ionizing radiation due to natural radioactivity in environment could be a precondition for spontaneous radiation mutagenesis in populations of any species.

The level of variability in a progeny of every species depends on the presence of mutagenic factors in the environment and the natural radioactive background can be realistically considered as one of possible reason for genetic variability from the beginning of evolution of species in the history of the Earth. Confirmation of a legitimacy of this assumption can be found in the fact of such sort: the known centers of origin of species are noted not only by conditions which assist a geographical isolation of new forms by also increased level of natural radioactivity.

The main mechanisms in the influence of the natural radioactivity as a mover of evolution can be related to increasing in the frequency of recombination in genetic structures, augmentation of intensity in error-prone DNA reparation and induction of genome instability.

Conclusively, the environment harbors large amounts of the natural radioactive elements in rocks, soils, waters, and atmosphere and of all kinds living beings. The background of the ionizing radiation is manifested everywhere on the Earth. The level of the background varies over a wide range of dose rates and modes of irradiation and is dependent on the content of the main natural primordial and cosmogenic radionuclides. The radioactive background has been kept at much higher level in the geological era when biochemical and cellular evolution have began. Species belonging to the immemorial taxones routinely accumulate the natural radioactive elements in abundance compared to comparatively recent species.

References

1. Danil'chenko, O.O. and Grodzinsky, D.M. (2004). Participation of DNA reparation systems in forming of radiation adaptive reaction in plants. In: "Paradigms of contemporary radiobiology. Radiation protection of personnel of the objects of atomic energetic. – Kiev: Chernobyl, pp. 15 – 16. (In Russian).
2. Eisenbud, M. (1987). Environmental Radioactivity. From Natural, Industrial and Military Sources. Third Edition, Academic Press, Inc., 475 p.
3. Gorev, L.M., Peleshenko, V.I. and Hil'chevsky, V.K. (1993). Radioactivity of the Natural Water. – Kiev: Vyshcha Shkola, 175 p. (In Ukrainian).
4. Grodzinsky, D.M. (1965). Natural radioactivity of plants and soils. –Kiev: Naukova dumka, 216 p. (In Russian).
5. Moiseev A.A. and Ivanov V.I. (1990). Manual on dosimetry and radiation health physics. Moscow: Energoatomizdat, 4th edition, 252p. (In Russian).
6. Perzov, L.A. (1964). Natural radioactivity of the Biosphere. Moscow: Atomizdat, 315 p. (In Russian).

7

Special Report on the Perceptions of Drug Abuse Research

S.K. Starling and Asheesh Kumar

Abstract

The past 30 years have seen a focus on substance abuse research in association with the creation of federal agencies specifically mandated to guide that effort. The purpose of this article is to provide a brief outline for the Perceptions of drug abuse. It contains different views like research involving the administration of abused drugs to humans, sexual & physical abuse and signs of drug abuse and dependence. This review is also focusing on specific drug classes like stimulants, sedatives, inhalants etc., their unpleasant effects and structures.

All things are poisons, for there is nothing without poisonous qualities. It is only the dose which makes a thing a poison.

Introduction

Drug abuse and addiction are major public health problems and contribute significantly to mortality and morbidity in our society (Musto, 1991; McGinnis and Foege, 1993; Wysowski, 1993). Together they cost society billions of dollars each year and contribute to considerable crime, personal suffering, and family misery (Rice *et al.*, 1990; Institute for Health Policy; 1993). At present, however, drug abuse treatment and prevention are not as effective as they could be and too many people are not being helped by existing interventions.

Clearly, we need to understand these problems better if we are to develop improved methods for treating those who are already addicted or for preventing new people from becoming drug abusers or drug dependent. Human drug research will play a vital role in the acquisition of this knowledge (Martin, 1995).

Research involving the administration of abused drugs to humans has a long and scientifically productive history (Wikler, 1953). It has been responsible for a number of important discoveries that have significantly improved our ability to influence the drug abuse problem. At the basic research level, it has enhanced our understanding of the behavioral, pharmacological, and physiological mechanisms that underlie drug abuse. These include mechanisms responsible for drug intoxication, tolerance, reinforcement, dependence and withdrawal.

Research participation also provides an opportunity for HIV/AIDS risk reduction. Approximately one-third of all AIDS cases in the United States are associated with drug abuse (Centers for Disease Control and Prevention, 1993). In addition to other health information, information on the dangers of needle sharing and unprotected sex are available to drug abusers in most research program. In many cases, HIV testing and counseling are also available. This information may reduce an addict's risk of contracting HIV infection or of transmitting that infection to others, including his/her sexual partner or children.

Histories of Sexual &Physical abuse are frequently reported by individuals participating in substance abuse treatment. Among women in drug treatment, alarmingly high rates of exposure to physical and sexual abuse are reported (Virginia *et al.*, 1997).

In the last decade, China experienced a rapid increase in illicit drug use, predominantly that of heroin. From 1990 to 2000, the number of drug abusers documented officially in China (e.g., those who were arrested for drug use by law reinforcement agencies or who voluntarily participated in drug treatment programs) increased 11-fold from 70,000 to 860,000 (UNAIDS, 2000).

According to the Regulations on Prohibition against Narcotics, approved by the Standing Committee of the National People's Congress in 1990, three levels of lawful penalty can be applied to illicit drug users. First-time drug users may be fined and/or placed in detention at local jails for up to 15 days. Those who use illicit drugs regularly and develop drug dependence are placed in residential compulsory rehabilitation centers for 1–6 months for detoxification (Yi-lang Tang *et al.*, 2005).

The person who takes the inventory indicates he has one of these problems:

- Health problems,
- Personality problems
- Social relationship problems
- Job related problems
- Problems associated with the misuse of leisure time
- Religious or spiritual problems

Signs of Drug Abuse and Dependence

The signs of drug abuse and dependence depend on the drug and the person using the drug. Behaviors that may be signs of a drug problem include:

- Changes in sleeping or eating habits, less attention to dressing and grooming, or less interest in sex.

Table 1

Problem Statement of different peoples who use illicit drugs regularly

S. No.	Problem Statement of different peoples
1.	I use alcohol and/or drugs to make bad feelings go away.
2.	When I'm out of work, I tend to drink more and/or use more drugs.
3.	Nearly all of my friends and acquaintances use drugs and/or drink heavily.
4.	It is very easy to get drugs in my neighborhood.
5.	When I get depressed or feel hurt, I just want to be left alone
6.	I spent most of my pay on alcohol and/or drugs.
7.	I used alcohol and/or drugs on the job and/or during lunch or break times.
8.	Rather than express anger, I'll go out and drink and/or use drugs.
9.	Other members of my family abuse alcohol and/or drugs.

- Up and down moods, a mood or attitude that is getting worse, or not caring about the future.
- Anger toward others or treating others badly.
- Sneaky behavior, lying, or stealing.
- Poor family relationships or relationships that are getting worse.
- New problems at work or school, or problems with the law.
- Dropping old friends and activities, finding new friends, and not wanting old friends to meet them.

Having these signs doesn't always mean a person is using drugs. The behavior could be because of work or school stress, or it could be a sign of depression or another medical problem. But behavior changes like these are common in people who abuse drugs.

Physical signs of drug abuse and dependence include:

- Red eyes, a sore throat, a dry cough, and feeling tired.
- Needle marks on the arm or other area of the body.
- Small, "pinpoint" pupils in the eyes.
- Losing weight without trying to, or not feeling like eating.
- Changes in sleep behavior, such as not sleeping as well.
- Seeing things that don't exist (hallucinations).

Drug Abuse Assessment- Reviews Focusing on Specific Drug Classes

Not surprisingly, many reviews on drug abuse assessment have been organized around specific classes of drugs of abuse. In this section, we discuss some of these.

Stimulants

("speed," "crack," "coke," "snow," "crank," "go," "speedball," "crystal," "cross-tops," "yellow jackets")

Cocaine: The abuse of cocaine increased dramatically in the late 1980s and early 1990s, but is now on the decline. Cocaine may be breathed in through the nose ("snorting"), or dissolved in water and taken through a vein (intravenously). When mixed with heroin for iv use, the combination is called a "speedball."

Through a simple chemical procedure, cocaine may be changed into a smokable form known as freebase or crack. Smoking produces an instant and intense sense of joy (euphoria), which is attractive to abusers. Other effects include:

- Feelings of increased confidence and energy
- Less inhibition
- Local numbness
- Powerful stimulation of the central nervous system

Increased use of and addiction to cocaine probably occur because it produces a very pleasurable high that is very short lived. This encourages the user to use the drug more often or regularly to get the desired effects.

Both the need to use larger amounts of the drugs to get the same effect (tolerance) and dependence may occur with regular cocaine use. Regular users may have:

- Depression
- Loss of interest in school, work, family, and friends
- Memory loss
- Mood swings
- Sleep problems
- Social withdrawal

Because heavy use may cause paranoia, cocaine users may become violent.

Amphetamines: During the 1950s and 1960s, amphetamines were often prescribed for conditions such as fatigue, obesity, and mild depression. Such use has stopped because the drugs are very addictive, and are now considered controlled substances.

Over-the-counter (OTC) amphetamine look-alike drugs are often abused. These drugs typically contain caffeine and other stimulants, and are sold as appetite suppressants or stay-awake/stay-alert aids.

If taken in high doses, these OTC drugs may cause the same high and other effects that occur with amphetamines. Regular users may have:

- Dilated pupils
- Irritability

- Restlessness
- Skin flushing
- Sleep disturbances
- Tremors
- Weight loss

Marijuana or Hashish

(also called "grass, pot, reefer, joint, hashish, cannabis, weed, and Mary Jane")

About 2 in 5 Americans have used marijuana at least once. About 10% of the population uses it on a regular basis. The source of marijuana is the hemp plant (cannabis sativa). The active ingredients are THC (delta-9-tetrahydrocannabinol) and other cannabinoids, which are found in the leaves and flowering shoots of the plant. Hashish is a substance taken from the tops of female plants. It contains the highest amount of THC. The drug dose in marijuana varies greatly depending on how it is prepared.

Cannabis is the most commonly used illegal drug in the United States (Johnston *et al.*, 2003), and research has consistently indicated that cannabis use is associated with an increased risk of other illegal drug use and problems (Donovan and Jessor, 1985; Hammer and Vaglum, 1990; Jessor, Chase and Donovan, 1980; Kandel *et al.*, 1986; Kandel *et.al.*, 1992; Morral, 2002). Furthermore, multiple studies have shown that the risk of substance abuse/dependence is greater among early illegal substance users than among later-users (Anthony and Petronis, 1995; Grant and Dawson, 1998; Hall and Lynskey, 2005; Lynskey *et al.*, 2003).

You may feel the effects of marijuana within seconds to several minutes after breathing in the smoke (from a joint or a pipe), or within 30 - 60 minutes after eating foods containing marijuana, such as "hash brownies." Because you can feel the effects almost right away, you can stop breathing in at any time to reduce the effect. In contrast, eating marijuana produces effects more slowly. These effects add up and last longer, making unpleasant reactions more likely.

The main effects of marijuana are on behavior, because the drug affects the central nervous system (CNS). Marijuana became popular because it gives people a feeling of joy (euphoria), relaxation, and increased sight, hearing, and taste with low to moderate doses. Most users also report an increase in their appetite ("the munchies").

Unpleasant effects that may occur include:

- Acute panic reactions or severe paranoia
- Changed body image
- Lack of orientation
- Trouble telling oneself from others

Inhalants

Inhalant use became popular with young teens in the 1960s with "glue sniffing." Since then, a greater variety of inhalants have become popular. Inhalant use typically involves younger teens or school-age children. Groups of children will use inhalants usually as an experiment. Commonly abused inhalants include:

- Aerosols for deodorants or hair sprays
- Cleaning fluids
- Gasoline
- Liquid typewriter correction fluid
- Model glue
- Spray paints

The chemicals are poured into a plastic bag or soaked into rags, then breathed in. The drugs are absorbed through the respiratory tract and cause an altered mental state within 5 - 15 minutes. Negative effects of inhalant abuse include:

- Brain damage
- Convulsion
- Liver or kidney damage
- Nerve damage (peripheral neuropathy)
- Sudden death

Most inhalant use occurs among teens or preteens who do not have access to illegal drugs or alcohol.

Opiates, opioids and Narcotics

Opiates come from opium poppies. These drugs include morphine and codeine. Opioids are artificial substances that have the same effect as morphine or codeine.

Opiates and opioids include:

- Heroin
- Hydromorphone
- Oxycodone
- Meperidine
- Methadone
- Propoxyphene

All of these substances, natural or human-made, are considered narcotics. Used as painkillers, these drugs:

- Change perception of painful stimuli
- Decrease anxiety

- Pinpoint pupils during intoxication
- Promote a relaxed state (sedation)

Because heroin is commonly injected into a vein (used intravenously), there are health concerns about sharing contaminated needles among iv drug users. Complications of sharing contaminated needles include hepatitis, HIV infection, and AIDS.

Hallucinogenic Drugs

Hallucinogenic agents, also called psychomimetic agents, are capable of producing hallucinations, sensory illusions and bizarre thoughts. The primary effect of these compounds is to consistently alter thought and sensory perceptions. Some of these drugs are used in medicine to produce model psychoses as aid in psychotherapy. Another purpose is to investigate the relationship of mind, brain, and biochemistry with the purpose of elucidating mental diseases such as schizophrenia.

A large body of evidence links the action of hallucinogenic agents to effects at serotonin receptor sites in the central nervous system. Whether the receptor site is stimulated or blocked is not exactly known. The serotonin receptor site may consist of three polar or ionic areas to complement the structure of serotonin as shown in the graphic (Figure 1.) on the left.

(a) ***Serotonin*:** Serotonin (5-hydroxytryptamine or 5-HT) is a monoamine neurotransmitter found in cardiovascular tissue, in endothelial cells, in blood cells, and in the central nervous system. The role of serotonin in neurological function is diverse, and there is little doubt that serotonin is an important CNS neurotransmitter.

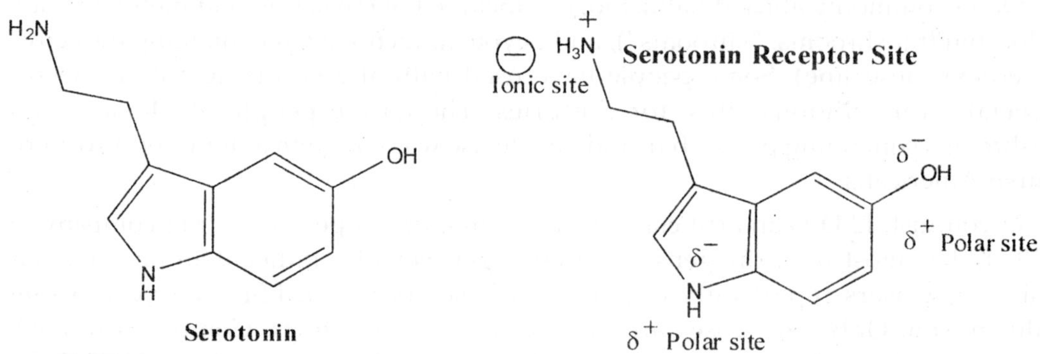

Figure 1. Structure of Serotonin and its Receptor Site

Although some of the serotonin is metabolized by monoamine oxidase, most of the serotonin released into the post-synaptic space is removed by the neuron through a re uptake mechanism inhibited by the tricyclic antidepressants and the newer, more selective antidepressant re uptake inhibitors such as fluoxetine and sertraline.

(b) *Mescalin and Psilocybin*: The drugs shown in the figure 2. can be isolated from natural sources: lysergic acid amide from morning glory seeds, psilocybin from the "magic mushroom", Psilocybe mexicana.

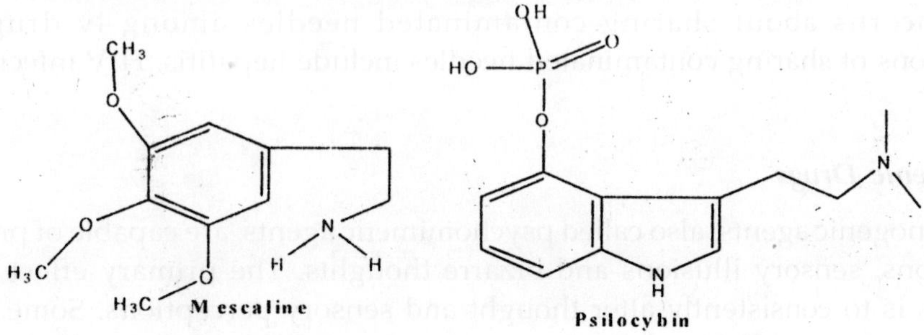

Figure 2. Structures of Mescalin and Psilocybin

The hallucinogenic molecules fit into the same receptors as the neuro-transmitter, and over-stimulate them, leading to false signals being created.

Mescaline is isolated from a peyote cactus. The natives of Central America first made use of these drugs in religious ceremonies, believing the vivid, colorful hallucinations had religious significance. The Aztecs even had professional mystics and prophets who achieved their inspiration by eating the mescaline-containing peyote cactus (Lophophora williamsii). Indeed, the cactus was so important to the Aztecs that they named it teo-nancacyl, or "God's Flesh". This plant was said to have been distributed to the guests at the coronation of Montezuma to make the ceremony seem even more spectacular.

Other commonly abused hallucinogens include LSD (lysergic acid diethylamide), psilocybin (mushrooms, "shrooms"), and peyote (a cactus plant containing the active ingredient mescaline). Some people have used naturally occurring hallucinogens, especially for religious rites, for centuries. The native people of Mexico used mushrooms containing psilocybin, and peyote use was common among southwestern Native Americans.

In contrast, LSD is an artificial substance, first developed by a drug company in 1938. Today, most hallucinogens are used experimentally rather than on a regular basis. Most users report only one or a few uses per year. LSD is a very strong hallucinogen. Only tiny doses are needed to produce effects. Compared to LSD, psilocybin is 100 - 200 times weaker, and mescaline (peyote) is about 4,000 times weaker.

Hallucinogens can lead to extreme anxiety and lack of reality at the height of the drug experience ("bad trips"). These experiences can come back as a "flashback," even without using the drug again. Such experiences typically occur during times of increased stress, and tend to occur less often and intensely after stopping the drugs.

(c) *LSD* (lysergic acid diethylamide): LSD is one of the most powerful hallucinogenic drugs known. (Figure 3.) LSD stimulates centers of the sympathetic nervous system in the midbrain, which leads to pupillary dilation, increase in body temperature, and rise in the blood-sugar level. LSD also has a serotonin-blocking effect.

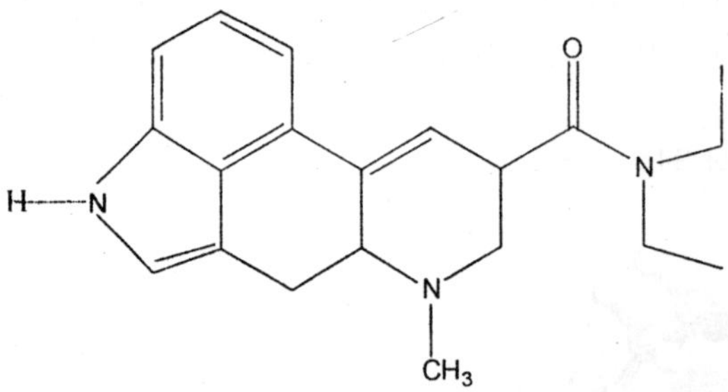

LSD- Lysergic Acid Diethylamide

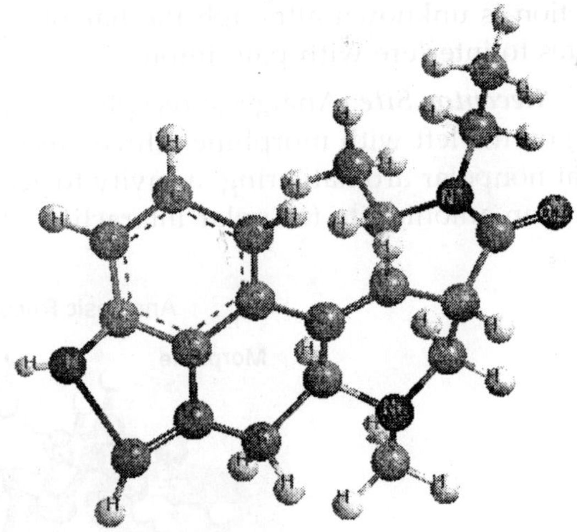

Figure 3. Structure of LSD (lysergic acid diethylamide)

The hallucinogenic effects of lysergic acid diethylamide (LSD) are also the result of the complex interactions of the drug with both the serotoninergic and dopaminergic systems.

During the first hour after ingestion, the user may experience visual changes with extreme changes in mood. The user may also suffer impaired depth and time perception, with distorted perception of the size and shape of objects, movements, color, sound, touch and the user's own body image.

Narcotic Analgesic Drugs

***Introduction*:** Narcotic agents are potent analgesics which are effective for the relief of severe pain. Analgesics are selective central nervous system depressants used to relieve pain. The term analgesic means "without pain". Even in therapeutic doses, narcotic analgesics can cause respiratory depression, nausea, and drowsiness. Long term administration produces tolerance, psychic, and physical dependence called addiction.

Narcotic agents may be classified into four categories:

1. Morphine and codeine - natural alkaloids of opium.
2. Synthetic derivatives of morphine such as heroin.
3. Synthetic agents which resemble the morphine structure.

4. Narcotic antagonists which are used as antidotes for overdoses of narcotic analgesics.

The main pharmacological action of analgesics is on the cerebrum and medulla of the central nervous system. Another effect is on the smooth muscle and glandular secretions of the respiratory and gastro-intestinal tract. The precise mechanism of action is unknown although the narcotics appear to interact with specific receptor sites to interfere with pain impulses.

Receptor Site: Analgesic receptor site may look as shown in the graphic (figure 4.) on the left with morphine. Three areas are needed: a flat area to accommodate a flat nonpolar aromatic ring, a cavity to accept another series of rings perpendicular, and an anionic site for polar interaction of the amine group.

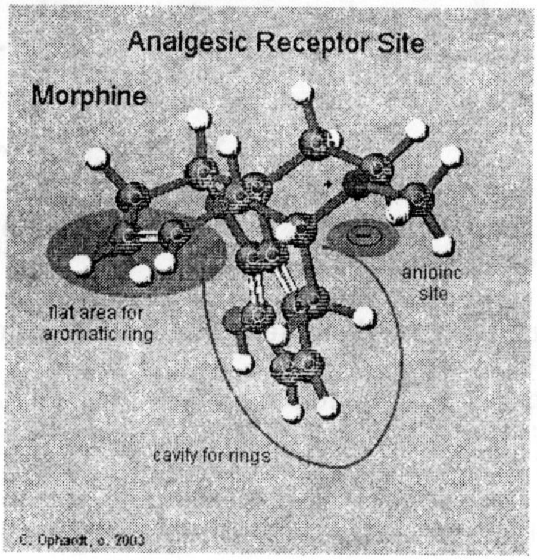

Figure 4. Structure showing the Receptor site in Morphine

Natural Peptide Analgesics – (a) *Enkephalins* : Recently investigators have discovered compound in the brain called enkephalins which resemble morphine in structure. Each one is a peptide composed of 5 amino acids and differs only in the last amino acid. The peptide sequences are: tyr-gly-gly-phe-leu and tyr-gly-gly-phe-met. Molecular models show that the structures of the enkephalins has some similarities with morphine. The main feature in common appears to be the aromatic ring with the -OH group attached (tyr). Methadone and other similar analgesics have two aromatic rings which would be similar to the enkephalins (tyr and phe).

Analgesics may relieve pain by preventing the release of acetylcholine. Enkephalin molecules are released from a nerve cell and bind to analgesic receptor sites on the nerve cell sending the impulse. The binding of enkephalin or morphine-like drugs changes the shape of the nerve sending the impulse in such a fashion as to prevent the cell from releasing acetylcholine. As a result, the pain impulse cannot be transmitted and the brain does not perceive pain.

(b) *Morphine and Codeine*: Morphine exerts a narcotic action manifested by analgesia, drowsiness, changes in mood, and mental clouding. The major medical action of morphine sought in the CNS is analgesia. Opiates suppress the "cough center" which is also located in the brainstem, the medulla. Such an action is thought to underlie the use of opiate narcotics as cough suppressants. Codeine appears to be particularly effective in this action and is widely used for this purpose.

Narcotic analgesics cause an addictive physical dependence. If the drug is discontinued, withdrawal symptoms are experienced. Although the reasons for addiction and withdrawal symptoms are not completely known, recent experiments have provided some information. A nucleotide known as cyclicadenosine monophosphate (cAMP) is synthesized with the aid of the enzyme adenylate cyclase. Enkephalin and morphine-like drugs inhibit this enzyme and thus decrease the amount of cAMP in the cells. In order to compensate for the decreased cAMP, the cells synthesize more enzymes in an attempt to produce more cAMP. Since more enzymes have been produced, more morphine is required as an inhibitor to keep the cAMP at a low level. This cycle repeats itself causing an increase in the tolerance level and increasing the amounts of morphine required. If morphine is suddenly withheld, withdrawal symptoms are probably caused by a high concentration of cAMP since the synthesizing enzyme, adenylate cyclase, is no longer being inhibited.

Morphine and codeine are contained in opium from the poppy (Papaver Somniterum) plant found in Turkey, Mexico, Southeast Asia, China, and India. This plant is 3-4 feet tall with 5-8 egg shaped capsules on top. Ten days after the poppy blooms in June, incisions are made in the capsules permitting a milky fluid to ooze out. The following day the gummy mass (now brown) is carefully scraped off and pressed into cakes of raw opium to dry. Opium contains over 20 compounds but only morphine (10%) and codeine (0.5%) are of any importance. Morphine is extracted from the opium and isolated in a relatively pure form. Since codeine is in such low concentration, it is synthesized from morphine by an ether-type methylation of an alcohol group. Codeine has only a fraction of the potency compared to morphine. It is used with aspirin and as a cough suppressant.

(c) *Heroin*: Heroin is synthesized from morphine by a relatively simple esterification reaction of two alcohol (phenol) groups with acetic anhydride (equivalent to acetic acid). Heroin is much more potent than morphine but without the respiratory depression effect. A possible reason may be that heroin passes the blood-brain barrier much more rapidly than morphine. Once in the brain, the heroin is hydrolyzed to morphine which is responsible for its activity. (Figure 5.)

Synthetic narcotic analgesics: may include the following:

Meperidine is the most common substitute for morphine. It exerts several pharmacological effects: analgesic, local anesthetic, and mild antihistamine. This multiple activity may be explained by its structural resemblance to morphine, atropine, and histamine.

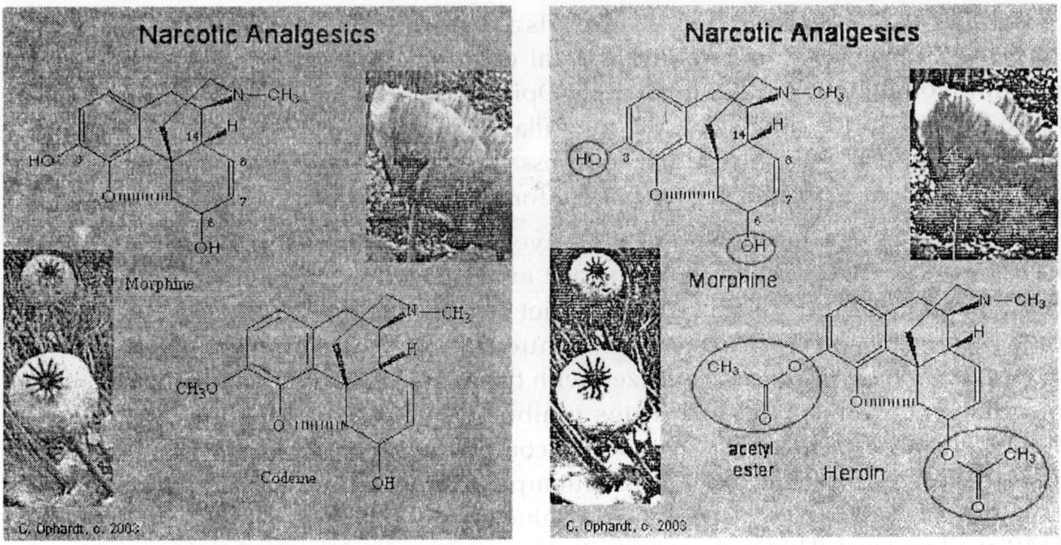

Figure 5. Structures of Morphine, Codeine and Heroin

(a) *Methadone*: Methadone is more active and more toxic than morphine. It can be used for the relief of may types of pain. In addition it is used as a narcotic substitute in addiction treatment because it prevents morphine abstinence syndrome.

Methadone was synthesized by German chemists during Wold War II when the United States and our allies cut off their opium supply. And it is difficult to fight a war without analgesics so the Germans went to work and synthesized a number of medications in use today, including demerol and darvon which is structurally simular to methadone. And before we go further lets clear up another myth. Methadone, or dolophine was not named after Adolf Hitler. The "dol" in dolophine comes from the latin root "dolor." The female name Dolores is derived from it and the term dol is used in pain research to measure pain e.g., one dol is 1 unit of pain.

Even methadone, which looks strikingly different from other opioid agonists, has steric forces which produce a configuration that closely resembles that of other opiates. (Figure 6).

In other words, steric forces bend the molecule of methadone into the correct configuration to fit into the opiate receptor.

When you take methadone it first must be metabolized in the liver to a product that your body can use. Excess methadone is also stored in the liver and blood stream and this is how methadone works its 'time release trick' and last for 24 hours or more. Once in the blood stream metabolized methadone is slowly passed to the brain when it is needed to fill opiate receptors. Methadone is the effective treatment for heroin addiction. It works to smooth the ups and down of heroin craving and allows the person to function nomrally.

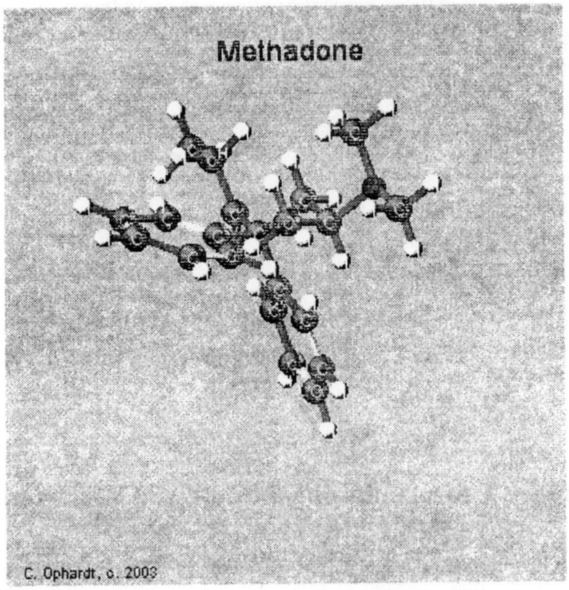

Figure 6. Structure of Methadone

(b) *Buprenorphine* : "Buprenorphine, an analgesic originally used iv for pre/post surgical analgesia and related, was approved not many years ago for opiate addiction treatment (brand names Subutex and, with naxolone added to help reduce iv abuse, Suboxone. It is provided as a sublingual tablet, available a many, but not all, pharmacies. As a practical matter, a big advantage over methadone is that it can be prescribed within a physician's office setting, not a "methadone clinic." The prescribing physician has to be certified to prescribe it, is issued a separate (2nd) DEA number, and is limited in the number of patients that may be seen for this treatment. Another advantage is that it does not have the abuse potential of methadone; it merely fills the opiate receptors to eliminate physical cravings. Onset of withdrawal from buprenorphine is slow, and the symptoms of withdrawal are generally milder, when compared with withdrawal from true opiates. It has been used in prison settings to effectuate a fairly painless withdrawal process in as little as 5 days, although in a typical setting the patient will be weaned off buprenorphine over a period of 4-8 weeks, and occasionally will continue on a maintenance dose."

Narcotic Antagonists: Narcotic Antagonists prevent or abolish excessive respiratory depression caused by the administration of morphine or related compounds. They act by competing for the same analgesic receptor sites. They are structurally related to morphine with the exception of the group attached to nitrogen.

Nalorphine precipitates withdrawal symptoms and produces behavioral disturbances in addition to the antogonism action. Naloxane is a pure antagonist with no morphine like effects. It blocks the euphoric effect of heroin when given before heroin. (Figure 7).

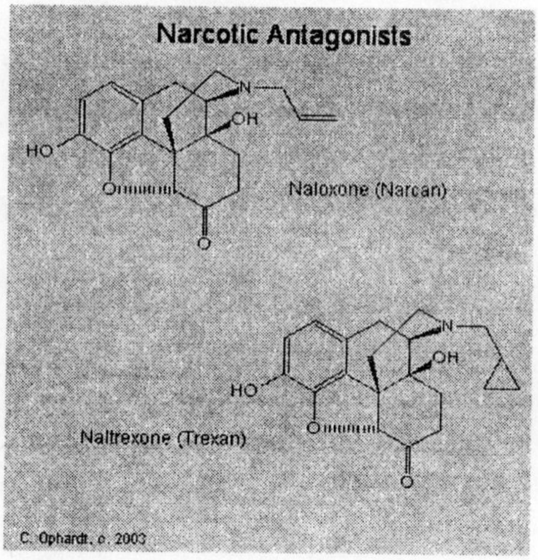

Figure 7. Structures of Naloxone and Naltrexone

Naltrexone became clinically available in 1985 as a new narcotic antagonist. Its actions resemble those of naloxone, but naltrexone is well is well absorbed orally and is long acting, necessitating only a dose of 50 to 100 mg. Therefore, it is useful in narcotic treatment programs where it is desired to maintain an individual on chronic therapy with a narcotic antagonist. In individuals taking naltrexone, subsequent injection of an opiate will produce little or no effect. Naltrexone appears to be particularly effective for the treatment of narcotic dependence in addicts who have more to gain by being drug-free rather than drug dependant.

Local Anesthetics

Introduction: Unlike other drugs which act in the region of the synapse, local anesthetics are agents that reversibly block the generation and conduction of nerve impulses along a nerve fiber. They depress impulses from sensory nerves of the skin, surfaces of mucosa, and muscles to the central nervous system. These agents are widely used in surgery, dentistry, and ophthalmology to block transmission of impulses in peripheral nerve endings.

Most local anesthetics can be represented by the following general formula. In both the official chemical name and the proprietary name, a local anesthetic drug can be recognized by the "-caine" ending.

The ester linkage can also be an amide linkage. The most recent research indicates that the local anesthetic binds to a phospholipid in the nerve membrane and inhibits the ability of the phospholipid to bind Ca^{+2} ions.

Practically all of the free-base forms of the drugs are liquids. For this reason most of these drugs are used as salts (chloride, sulfate, etc.) which are water soluble, odorless, and crystalline solids. As esters these drugs are easily hydrolyzed with consequent loss of activity. The amide form of the drug is more stable and resistant to hydrolysis.

(a) *Benzocaine and Lidocaine*: Two local anesthetics are shown in the figure 8.

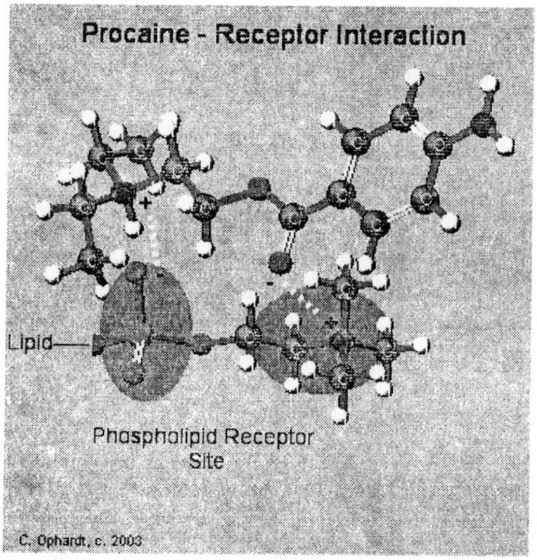

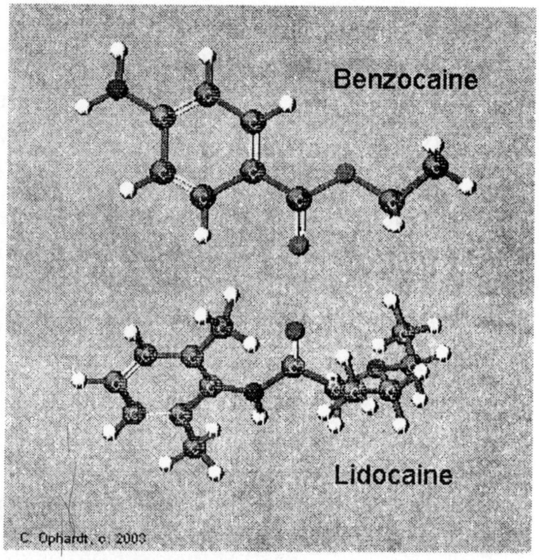

Figure 8. Structures of Benzocaine and Lidocaine and Receptor interaction in Procaine

Central Nervous System

Drugs Acting Upon the Central Nervous System

The central nervous system directs the functions of all tissues of the body. The peripheral nervous system receives thousands of sensory inputs and transmits them to the brain via the spinal cord. The brain processes this incoming information and discards 99% as unimportant. After sensory information has been evaluated, selected areas of the central nervous system initiate nerve impulses to organs or tissue to make an appropriate response.

Chemical influences are capable of producing a myriad of effects on the activity and function of the central nervous system. Since our knowledge of different regions of brain function and the neurotransmitters in the brain is limited, the explanations for the mechanisms of drug action may be vague. The known neurotransmitters are: acetylcholine which is involved with memory and learning; norepinephrine which is involved with mania-depression and emotions; and serotonin which is involved with biological rhythms, sleep, emotion, and pain.

Central Nervous System Stimulants:

Stimulants are drugs that exert their action through excitation of the central nervous system. Psychic stimulants include caffeine, cocaine, and various amphetamines. These drugs are used to enhance mental alertness and reduce drowsiness and fatigue. However, increasing the dosage of caffeine above 200 mg (about 2 cups of coffee) does not increase mental performance but may increase nervousness, irritability, tremors, and headache. Heavy coffee drinkers become psychically dependent upon caffeine. If caffeine is withheld, a person may experience mild withdrawal symptoms characterized by irritability, nervousness, and headache.

(a) *Xanthines* : Caffeine and the chemically related xanthines, theophylline and theobromine, decrease in the order given in their stimulatory action. They may be included in some over-the-counter drugs.

The action of caffeine is to block adenosine receptors as an antagonist. As caffeine has a similar structure to the adenosine group. This means that caffeine will fit adenosine receptors as well as adenosine itself. It inhibits the release of neurotransmitters from presynaptic sites but works in concert with norepinephrine or angiotensin to augment their actions. Antagonism of adenosine receptors by caffeine would appear to promote neurotransmitter release, thus explaining the stimulatory effects of caffeine.

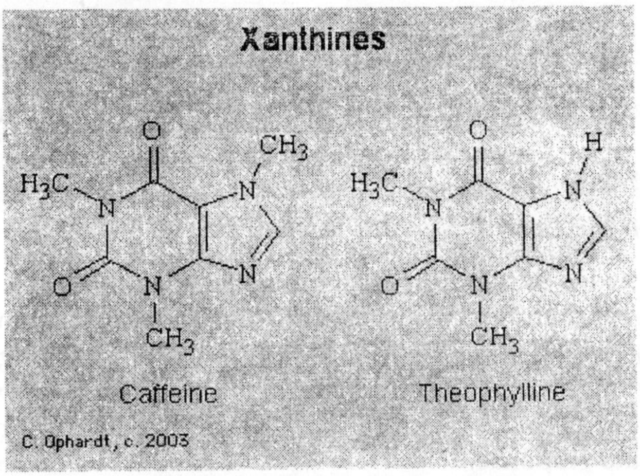

Figure 9. Structures of xanthenes

(b) *Amphetamines:* The stimulation caused by amphetamines is caused by excessive release of norepinephrine from storage sites in the peripheral nervous system. It is not known whether the same action occurs in the central nervous system. Two other theories for their action are that they are degraded slower than norepinephrine or that they could act on serotonin receptor sites.

Therapeutic doses of amphetamine elevate mood, reduce feelings of fatigue and hunger, facilitate powers of concentration, and increase the desire and capacity to carry out work. They induce exhilarating feelings of power, strength, energy, self-assertion, focus and enhanced motivation. The need to sleep or eat is diminished.

Levoamphetamine (Benzedrine), dextroamphetamine (Dexedrine), and methamphetamine (Methedrine) are collectively referred to as amphetamines. (Figure 10).

Benzedrine is a mixture of both the dextro and levoamphetamine isomers. The dextro isomer is several times more potent than the levo isomer. The misuse and abuse of amphetamines is a significant problem which may include the house wife taking diet pills, athletes desiring an improved performance, the truck driver driving non-stop coast to-coast, or a student cramming all night for an exam.

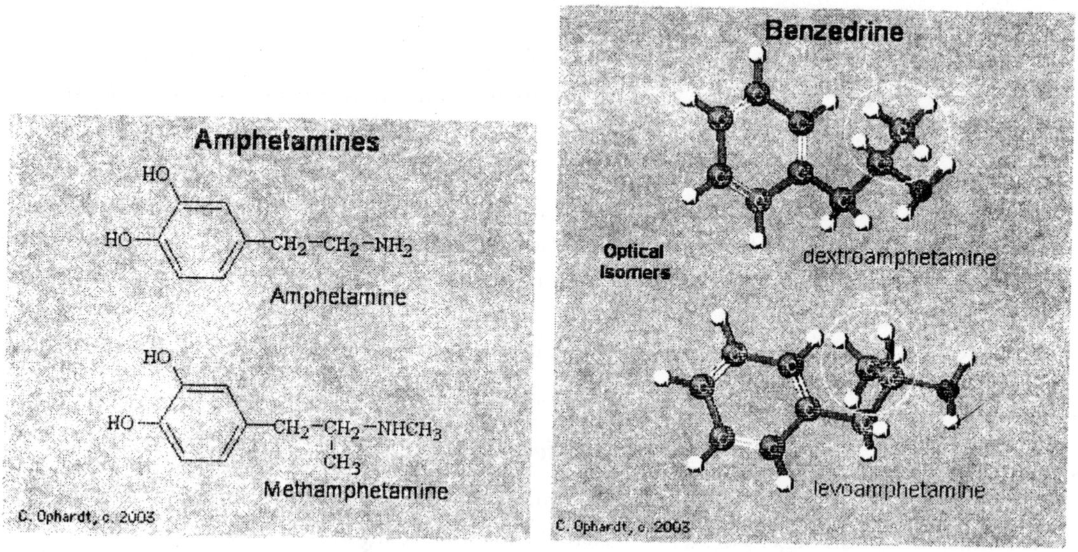

Figure 10. Structures of Amphetamines

Conclusion

Drug abuse is rife in many countries. Billions of dollars are spent internationally preventing drug use, treating addicts, and fighting drug-related crime. Although drugs threaten many societies, their effects can also be combated successfully. This essay looks at some of the effects of drug use on society, and suggests some solutions to the problem.

Drug abuse is a complex problem thought to result from a combination of hereditary, psychological, and environmental factors. It affects people from the neonatal stage to old age. Infants of abusers may suffer from neglect or the effects of parental drug use. As they grow into childhood, they may demonstrate antisocial behavior, and signs of malnutrition, poor self-esteem, depression, or attention deficit disorder. This may lead an adolescent to use drugs, have unwanted pregnancies, and drop out of school.

Addicts cannot function as normal members of society. They neglect or abuse their families, and eventually require expensive treatment or hospitalization. The second effect is on crime. Huge police resources are needed to fight smuggling and dealing. Criminal gangs and mafia underworlds develop with the money from drugs.

However, the menace of drugs can be fought. Education is the first battle. Children need to be told at home and in school about drugs. People need to be aware of the effects so that they can make avoid this problem. A second approach is to increase police manpower and powers to stop dealers and to enforce the law. However the main target should be the user. Families and counselors need to talk to children and people at risk. Parents need to look at their children and help them to Jobs are needed to give people a role in society.

In conclusion, although the problem of drugs may seem impossible to eliminate, there are concrete steps that can be taken to weaken the hold of drugs on society. The danger from drugs is too great to ignore.

Much progress has been made in understanding the relationship between substance abuse and the family. Family factors have an important role in explaining the onset, development, and amelioration of substance abuse. Encouraging gains have been made in treating couples and families with substance abuse problems. More research is needed involving diverse and international populations, but efficacious treatments are growing in type and number that bodes well for improving the lives of millions worldwide.

References

Musto, D.F. (1991). Opium, cocaine and marijuana in American history. *Scientific Am.* 265, 40-47.

McGinnis, J.M. and Foege, W.H. (1993). Actual causes of death in the United States. *Journal of the American Medical Association*, 270, 2207-2212.

Wysowski, D.K., Schober, S.E., Wise, R.P. and Kopstein, A. (1993). Mortality attributed to misuse of psychoactive drugs, 1979-1988, *Public Health Rep*. 108, 565-570.

Rice, D.P., Kehnan, S., Miller, L.S. and Dunmeyer, S. (1990). The Economic Costs of Alcohol and Drug Abuse and Mental Illness (1985), DHHS Publication No. (ADM) 90-1694, San Francisco, CA.

Institute for Health Policy (1993). Substance Abuse: The Nation's Number One Health Problem, The Robert Wood Johnson Foundation, Princeton, NJ.

Martin W. Adler (1995). College on Problems of Drug Dependence, *Drug Alcohol Depend*. 37, 167-175.

Wikler, A. (1953). Opiate Addiction, Psychological and Neurophysiological Aspects in Relation to Clinical Problems, Charles C. Thomas, Springfield, IL.

Centers for Disease Control and Prevention (1993). *HIV/AIDS Surveillance Report*, Vol. 5, No.2.

Virginia G.R., Ma, Robert F., M. Douglas A., and Ellise T., Ba, (1997). *Journal of Substance Abuse Treatment*, 14, 351-358.

UNAIDS (2000). Drug use and HIV vulnerability: Policy research study in Asia (pp. 37 - 52). Bangkok, Thailand 7 UNAIDS Asia Pacific Intercountry Team and UNDCP Regional Centre for East Asia and the Pacific

Yi-lang Tang, T. Anna Wiste, Pei-xian Mao, Ye-zhi Hou, (2005). *Journal of Substance Abuse Treatment,* 29, 215– 220.

Johnston, L.D., O'Malley, P.M., & Bachman, J. G. (2003). *Monitoring the Future national survey results on drug use,* 1975–2002. Bethesda, MD: National Institute on Drug Abuse.

Donovan, J.E., & Jessor, R. (1985). Structure of problem behavior in adolescence and young adulthood. *Journal of Consulting and Clinical Psychology,* 53, 890-904.

Hammer, T. & Vaglum, P. (1990). Use of alcohol and drugs in the transitional phase from adolescence to young adulthood. *Journal of Adolescence,* 13, 129-142.

Jessor, R., Chase, J.A., & Donovan, J.E. (1980). Psychosocial correlates of marijuana use and problem drinking in a national sample of adolescents. *American Journal of Public Health,* 70, 604-610.

Kandel, D.B., Davies, M., Karus, D. & Yamaguchi, K. (1986). The consequences in young adulthoodof adolescent drug involvement: An overview. *Archives of General Psychiatry,* 43, 746-754.

Kandel, D.B., Yamaguchi, K. & Chen, K. (1992). Stages of progression in drug involvement from adolescence to adulthood: Further evidence for the gateway theory. *Journal of Studies on Alcohol,* 53, 447-457.

Morral, A.R., McCaffrey, D.F., & Paddock, S.M. (2002). Reassessing the marijuana gateway effect. *Addiction,* 97, 1493-1504.

Anthony, J.C., & Petronis, K.R. (1995). Early-onset drug use and risk of later drug problems. *Drug and Alcohol Dependence,* 40, 9-15.

Grant, B.F., & Dawson, D.A. (1998). Age of onset of drug use and its association with DSM-IV drug abuse and dependence: Results from the National Longitudinal Alcohol Epidemiologic Survey. *Journal of Substance Abuse,* 10, 163-173.

Hall, W.D., & Lynskey, M. (2005). Is cannabis a gateway drug? Testing hypotheses about the relationship between cannabis use and the use of other illicit drugs. *Drug and Alcohol Review,* 24, 39-48.

Lynskey, M.T., Heath, A.C., Bucholz, K.K., Slutske, W.S., Madden, P.A.F., Nelson, E. C. (2003). Escalation of drug use in early-onset cannabis users vs. co-twin controls. *Journal of the American Medical Association,* 289, 427-433.

8

Phytoremediation of Heavy Metals: Potential of Indigenous Species

M.S. Paul, Rohan D'Souza and Mayank Varun

Introduction

Man, as a species has surpassed and dominated all other life forms on this planet not only because of his inherent ability to better adapt himself to his environment, but also due to his genius in modifying and exploiting his immediate surroundings for his benefit. Ever since the Industrial revolution that changed the world in 1900, these activities have gone ahead at a mindboggling pace, too fast in fact, to keep an eye on. Unfortunately, in the absence of regulations and monitoring, anthropogenic activities have seriously and steadily altered the natural biogeochemical cycles of the environment. All the realms of the biosphere—water, air and soil—are reeling under the impact of pollution.

HEAVY METAL CONTAMINATION OF SOIL

Soil is one of the natural resources being overexploited globally due to increased industrial, agricultural and other human activities. Soil contamination, both diffuse and localized, can lead to severe damage of soil functions as well as contamination of surface and groundwater. Urban soils are increasingly acting as a sink for a wide range of contaminants including heavy metals due to the rapid pace of development. Heavy metals are a threat to urban environmental quality and health of urban residents as they persist in the pedosphere for a long period of time, even after removing the pollution sources (Imperato *et al.*, 2003). Contamination sources include mining and refining of metals, sludge dumping, energy and fuel production, power transmission, intensive agriculture, etc. (Nouri *et al.*, 2009; Samarghandi *et al.*, 2007).Vehicular emissions are also a significant source of many pollutants, as reported (Viard *et al.*, 2004). Stringent standards for maximum permissible limits of heavy metals have been set by agencies like the World Health Organization (WHO) and the United states Environment Protection Agency (US EPA). In the past few years, heavy metal contamination of soils and water from industrial and traffic sources in urban environments has been studied in North America and Europe (Crnkoviæ *et al.*, 2006, Madrid *et al.*, 2004; Chirenje *et al.*, 2004; Imperato *et al.*, 2003 and Tijhuis, 2002).

Toxic Effects of Heavy Metals

Although they do not occur naturally in living organisms and have been reported not to have any known function in them (Lenntech, 2004), heavy metals are absorbed by and tend to accumulate in humans and animals through food, air and water. Lead, cadmium, copper, zinc, nickel, chromium and arsenic are the metals most frequently reported to have the highest impact on organisms (Vamerali *et al.*, 2010; Liphadzi and Kirkham, 2005). Although individual metals exhibit specific signs of their toxicity, the following have been reported as general signs associated with cadmium, lead, arsenic, mercury, zinc, copper and aluminium poisoning: gastrointestinal disorders, diarrhoea, stomatitis, tremor, ataxia, paralysis, vomiting and convulsion, depression, and pneumonia when volatile vapours and fumes are inhaled (McCluggage, 1991). The nature of effects could be toxic (acute, chronic or sub-chronic), neurotoxic, carcinogenic, mutagenic or teratogenic.

Lead is the most significant toxin of the heavy metals, and the inorganic forms are absorbed through ingestion by food and water, and inhalation (Ferner, 2001). A serious effect of lead toxicity is its teratogenic effect. Lead poisoning also causes inhibition of the synthesis of haemoglobin; dysfunctions in the kidneys, joints and reproductive systems, cardiovascular system, gastrointestinal tract and acute and chronic damage to the central and peripheral nervous system (INECAR, 2000). Cadmium is toxic at extremely low levels. In humans, long term exposure results in renal dysfunction and lung diseases. Cadmium is also associated with bone defects, viz; osteomalacia, osteoporosis and spontaneous fractures, increased blood pressure and myocardic dysfunctions (INECAR, 2000). Zinc has been reported to cause the same signs of illness as does lead, and can easily be mistakenly diagnosed as lead poisoning (McCluggage, 1991). Zinc is considered to be relatively non-toxic, especially if taken orally. However, excess amount can cause system dysfunctions that result in impairment of growth and reproduction (INECAR, 2000; Nolan, 2003). Inorganic forms of mercury cause spontaneous abortion, congenital malformation and gastrointestinal disorders. Poisoning by its organic forms causes severe skin diseases, gingivitis, stomatitis, neurological disorders, total damage to the brain and CNS and are also associated with congenital malformation (Ferner, 2001; Lenntech, 2004).As with lead and mercury, arsenic toxicity symptoms depend on the chemical form ingested (Ferner, 2001). Arsenic acts to coagulate protein, forms complexes with coenzymes and inhibits the production of adenosine triphosphate (ATP) during respiration (INECAR, 2000). It is possibly carcinogenic and high-level exposure can cause death (USDOL, 2004). Ingestion of elevated levels of Cu causes gastrointestinal distress, while long-term exposure to high Cu concentration causes liver and kidney damage (USEPA 2002). Ni toxicity inhibits spermatogenesis, amylase enzymes, insulin formation and kidney function (Srivastava and Gupta 1996). Effects of Mn toxicity on animals and humans are essentially not known (Liphadzi and Kirkham, 2005).

Conventional Control Measures

Several technologies exist for remediation of metal contaminated soils. Conventional remediation technologies to clean polluted soils include soil incineration, excavation and landfill, soil washing, soil flushing, solidification and stabilization electro kinetic systems. Each of these technologies has specific benefits and limitations (EPA, 1997). Immobilization and extraction by physico-chemical methods can be expensive; in effect, limiting their suitability for areas where rapid complete decontamination is the prerequisite (BIO-WISE, 2000). Also, remediation of soil contamination by conventional engineering techniques often costs between $50-$500 per ton. Customized techniques developed for individual sites can even cost over $1000 per ton. For an acre of soil to depth of 3 ft. i.e. approx. 4500 tons of soil, the figure works out to be a minimum of a quarter million dollars (Cunningham and Ow, 1996). These figures have limited the widespread and ready remediation of contaminated sites. Moreover, some techniques like soil washing have an adverse effect on biological activity, soil structure and fertility (Datta and Sarkar, 2004). Thus, the task before scientists today is to find *in-situ* processes that can be used to remediate such soils inexpensively.

Phytoremediation

The global emphasis at present is to use natural materials and methods as far as possible to combat a variety of problems. The same has been extended to soil remediation programmes. A variety of technologies that employ biological resources to clean up metal polluted sites have gained considerable momentum in the last decade or so, both in developed as well as in developing nations, and are currently in the process of commercialization (Prasad, 2004).

It is known that plants can cause profound alteration of surrounding soil by the simple life processes like water and mineral uptake, decay etc. Phytoremediation is a practical, natural method that relies on plants to extract and concentrate pollutants from the environment into their tissues (Raskin and Salt, 1997). It is a feasible and practical 'natural' remediation technique for contaminated soils and may be defined as "the engineered use of green plants to remove, contain or render harmless environmental contaminants like heavy metals, trace elements, organic compounds and radioactive compounds in soil or water" (Hinchman, *et al.*, 1998). The various pathways of soil-metal-plant interaction have been elucidated in Table 1 and Figure1.

Phytoremediation is especially suited for countries where labour, expertise and cost saving techniques are vital due to constraints of funds for other methods. As an emerging tool for *in-situ* remediation to clean up contaminated soils and water, phytoremediation has gained attention because it is potentially cheaper; approx.

Table 1

Overview of Phytoremediation processes

Mechanism	definition	Media	Contaminants	Reported Plants	Status of research
Phytoextraction	Plants remove pollutants from soil and concentrate them in harvestable parts	Soil, sediment, sludges	Inorganics: Metals: Ag, Cd, Co, Cr, Cu, Hg, Mn, Mo, Ni, Pb, Zn. Radionuclides: Sr, ^{137}Cs, ^{230}Pb, $^{238,234}U$	Indian mustard, alyssum, hybrid poplars pennycress, sunflower,	Laboratory, pilot and field applications
Rhizofiltration	Plant roots absorb pollutants from water	Groundwater, surface water	Organics/Inorganics: Metals, radionuclides	Sunflower, Indian mustard, water hyacinth	Laboratory, pilot scale
Phytostabilization	Plants reduce mobility and bioavailability of pollutants by immoblisation	Soil, sediment, sludges	Inorganics: As, Cd, Cr, Cu, Hs,Pb, Zn	Indian mustard, hybrid poplars, grasses	Field application
Phytodegradation	Plants and associated microbes degrade pollutants	Soil, sediment, sludges, groundwater surface water	Organic compounds (TPH, PAHs, pesticides, PCBs, chlorinated solvents,)	Red mulberry, grasses, hybrid poplars, black willow, bald cypress cattail, rice, algae	Field demonstration/ application
Phytovolatilization	Plants volatilize pollutants from media into the atmosphere	Soil, sediment, sludges, groundwater	Organics/Inorganics: Chlorinated solvents, some inorganics (Se, Hg, As)	Poplars, alfalfa, black locust, Indian mustard	Laboratory and field application

(Adapted from Erakhrumen, 2007)

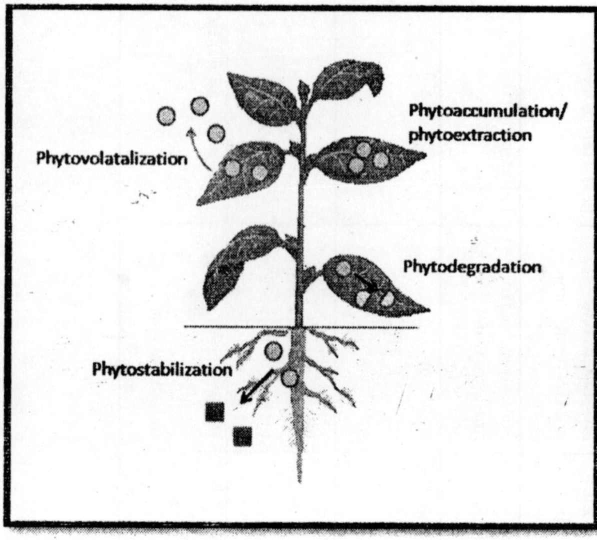

Figure 1: Pathways of Phytoremediation

90% (Hinchman, *et al.*, 1998) than conventional treatments, which are primarily chemical based and pursue an energy intensive approach. It leaves the topsoil in usable condition and it is aesthetically pleasing. It requires minimal equipment and less energy inputs as plants do most of the work using solar energy. Thus, it is an eco-friendly process. The plants used can later be harvested, processed and disposed off in an environmentally sound manner. This approach is hence emerging as an innovative tool with a great potential to decontaminate soil and water and thus achieving a sustainable development status (Desouza *et al.*, 1999).

Phytoextraction is one of the strategies of phytoremediation, where plants are used to uptake and accumulate heavy metals in above ground biomass, which can be harvested and removed from the site. Phytoextraction efficiency is determined by two key factors: metal hyperaccumulating capacity and biomass production (McGrath and Zhao, 2003). Depending upon the trace element accumulation capability, plants have been identified as excluders (with avoidance mechanisms for element uptake), highly sensitive indicators (lacking protection mechanisms) and accumulators with mechanisms of metal tolerance and accumulation capability in above ground biomass (Baker *et al.*, 2000). Phytostabilisation, another phytoremedial technique, aims to reduce the mobility and bioavailability of pollutants in environment and not removing them (Dietz and Schnoor, 2001). The establishment of a green canopy in polluted soil has the effect of reducing the mobility of pollutants through water, wind erosion and water percolation. A significant fraction of metals can be stored at root level, especially in polyannual species (Vamerali *et al.*, 2009) contributing to long-term stabilisation of pollutants.

Hyperaccumulators v/s Nonhyperaccumulating Wild Species

Plants show differing morpho-physiological responses to soil metal contamination. Most are sensitive to very low concentrations; others have developed tolerance, and a reduced number show. The latter capacity has practically opened up the way to phytoextraction. Hyperaccumulation is an unusual occurrence, seen in a narrow range of species which often grow in metal-rich soils. By analysing the metal contents of several species from worldwide collections, Baker and Brooks (1989) suggested the following values of metal concentrations in shoots for hyperaccumulation without evident symptoms of toxicity: 100 mg kg^{-1} for Cd, 1,000 mg kg^{-1} for Ni, Cu, Co and Pb, and 10,000 mg kg^{-1} for Zn and Mn (Table 2). Although these threshold values were fixed arbitrarily, they roughly lie at least at one order of magnitude greater than those found in common species (Salt and Kramer 2000). The known hyperaccumulators are generally minor vegetation components in most European and North-American habitats. Currently, more than 400 hyperaccumulator species are known, belonging to 45 different botanical families, among which the most frequent are Brassicaceae and Fabaceae (Baker *et al.*, 2000). Among monocotyledons, grasses hold promise as hyperaccumulators (Prasad and De Oliveira-Freitas, 2003). Lack of information on the agricultural management of hyperaccumulators, together with slow-growing and poor shoot and root growth, increase the difficulties in the practical application of these species in remediation projects (Lasat, 2002).

The choice of plant species is an important task in any phytoremediation based technique. Decontaminating a site in a reasonable number of harvests requires plants that are both high yielders of biomass and good metal accumulators by dry weight. It has been demonstrated (Porêbska and Ostrowska, 1999; Saraswat and Rai, 2009; Nedunuri *et al.*, 2010) that, wild native plants may be better phytoremediators for waste lands than the known metal bioaccumulators from literature. Although known metal hyperaccumulators like *Thlaspi caerulescens* and *Alyssum bertolonii* are able to take up and accumulate appreciable amounts of metals in their tissues, their use in the field is limited because these are slow growing with shallow root systems, and are not very widespread. In most cases their biomass yield is also quite low (Saraswat and Rai, 2009; Saifullah *et al.*, 2009). Also, the technology for their large-scale cultivation is not fully developed; therefore, their use is rather limited (Pulford and Watson, 2003). Generally, metal toxicity issues do not arise in natural soils with their native vegetation. Even if the soil is naturally high in a particular metal, native plants will often have become adapted over time to the locally elevated levels (Brooks *et al.*, 1992; Ouzounidou *et al.*, 1994). Allowing native species to remediate soils is an attractive proposition since native wild species do not require frequent irrigation, fertilization, and pesticide treatments, while simultaneously a plant community comparable to that existing in the vicinity can be established. The outcome is, thus, both site remediation and ecological restoration (Henry, 2006). A range of native,

Table 2
Heavy metals: their role and impact

Metal	Essential for Plants	Functions in plants	Some hyperaccumulators	Hyper-accumulating threshold (mg kg^{-1} dw)	Toxicity symptoms in humans
As	No	-	*Pteris vittata*	>1,000	Cancer (e.g. lung and skin); cardiovascular, gastrointestinal, hepatic and renal diseases
Cd	No	-	*Thlaspi caerulescens*	>100	Stomach irritations (vomiting and diarrhoea); lung damage; kidney diseases; cancer (probably)
Co	Yes	Cofactor of biosynthetic enzymatic activities; essential for Rhizobium	*Haumaniastrum robertii*	> 1,000	Contact dermatitis; mutagenic and carcinogenic effects
Cr	No	–	*Brassica juncea.; B. napus Vallisneria americana*	> 20,000	Genotoxic carcinogens; lung cancer
Cu	Yes	Constituent of enzymes; role in photosynthesis; involved in reproduction and determining yield and quality in crops	*Brassica juncea.; Vallisneria americana*	> 1,000	Inhibition of haemopoiesis; accumulation in liver and kidney
Mn	Yes	Constituent of enzymes; activation of enzymes; photosynthesis; reproductive phase; resistance against biotic and abiotic stress	*Agrostis castellana*	> 10,000	Neurological symptoms; affection of liver Function
Ni	Yes	Constituent of enzymes; activation of urease	*Alyssum spp.; Thlaspi spp.*	> 1,000	Accumulation in lungs
Pb	No	-	*Thlaspi rotundifolium ; T. caerulescens.; Alyssum wulfenienum Arrhenatherum elatius ; Festuca ovina*	> 1,000	Irreversible neurological damage; renal disease; cardiovascular effects; reproductive toxicity
Zn	Yes	Constituent of cell membranes; activation of enzymes; DNA transcription; involved in reproductive phase and in determining yield and quality of crops; resistance against biotic and abiotic stress; legume nodulation and nitrogen fixation	*Thlaspi spp.; Cardaminopsis spp.*	> 10,000	Inhibition of copper absorption; nausea, vomiting, loss of appetite, abdominal cramps, diarrhoea, headaches

Adapted from Vameralli *et al.*, 2010.

well adapted plants have been investigated and even used for heavy metal bioindicatoring and phytoremedial purposes including lemongrass and other wild grasses, vetiver, *Sesbania, Avena, Crotalaria, Crinum asiaticum* and *Calotropis procera* etc.(Yang *et al.*, 2003; Uraguchi *et al.*, 2006; Varun *et al.*, 2010; D'Souza *et al.*, 2010).

Figure 2. Plant species with phytoremedial potential (a) *Calotropis procera* [Cd, Cu]; (b) *Chenopodium murale* [Cd, Cu]; (c) *Crinum asiaticum* [Cd, Pb, Cu]; (d) *Poa annua* [Pb, Cu].

Bioavailability of Heavy Metals

Pollutant bioavailability depends on the chemical properties of the pollutant, soil properties, environmental conditions and biological activity. In general soil pH seems to have the greatest effect of any single factor on the availability of metals in soil as the equilibrium between metal speciation, solubility, adsorption and exchange is intimately connected to soil pH (Sauve *et al.*, 1997). In accordance with the changes in metal bioavailability associated with a change in pH, many studies have found that plant uptake of Mn and Zn increases as soil pH decreases (Turner, 1994; Sauve *et al.*, 1997).

Organic matter forms metal complexes, so that it can either reduce metal mobility or increase availability when the complexes are soluble in the soil solution (Halim *et al.*, 2003). Generally, plants being unable to absorb the large metal-complexes, the bioavailability of metals decreases. Cu is often predominantly found bound to the organic matter fraction in the soil and this can be the most important soil factor in determining Cu bioavailability. The amount of organic matter found in soils also affects the bioavailability of Zn and Mn (del Castilho *et al.*, 1993), though not in a major way. Metals like Pb are mostly immobile in the soil, which reduces their bioavailability (Lombi *et al.*, 2001) and subsequent uptake by the plant, Cd on the other hand is usually less adsorbed by soil and organic matter which makes it more available to plants (McLaughlin *et al.*, 2000), even though it is far more toxic. Conversely, the anionic behaviour of As leads to its greater mobility in conditions of high pH and organic matter (Hartley *et al.*, 2009).

Mineral nutrition profoundly influences the growth of plants and the absorption of nutrients by plant roots, two areas with which phytoextraction is greatly concerned. Proper plant nutrition has the potential to be an effective, low-cost agronomic practice for enhancing the phytoextraction of heavy metals by plants, but more research is required before fertilizers can be used effectively for this purpose. (Hamlin, 2002). Chelates like EDTA, CDTA, HEDTA etc and biosurfactants can also be used to increase the bioavailability of metal ions in the rhizosphere. These supplements increase the mobility of metal ions in the soil. (Cunningham & Ow, 1996)

Potential of Indigenous Species

On evaluating the phytoremediation potential of native flora in Agra district covering some of the industrial, residential and agricultural sites a comprehensive profile of Co, Cd, Pb and Cu contamination was obtained. Cd exhibited a greater accumulation in all soils, whereas, accumulation of Cu was high in limited samples. A total of 28 plants species were evaluated in which 21 herbs, 5shrubs and 2 grasses were assessed.

Individual elements displayed a remarkably different pattern of accumulation inn soils as well as plants. Plants established limited capabilities in mobilizing Co and Pb in the root zone. Cu was predominantly prohibited in shoots; Co and Cd in roots while Pb was almost equally between shoots and roots. Chlorophyll and free proline pattern indicate that most plants have evolved strategies to counter phytotoxicity due to accumulated Cd, Pb and Cu at much above reported toxic levels.

Heavy metal content observed in plant roots and shoots is presented in table 3. On studying the trends of accumulation and translocation of individual elements, each species indicates varying uptake patterns. Translocation patterns of Cu indicate that it was mainly partitioned in the shoots (by 64% plants). Pb was almost equally partitioned between shoots and roots. Co and Cd were predominantly partitioned in the roots with 68% and 64% plants, respectively.

Pb is not easily transferred to above-ground plant biomass, being stored mainly in root cells (Mellem *et al.*, 2009). As far as the thresholds for accumulation in aboveground biomass for different heavy metals achieved by some hyperaccumulators (Vamerali *et al.*, 2010) are concerned, none of the species displayed such idealistic values except for the grass *Poa annua* which at 742.06 mg/kg dry weight in shoot was able to come near the threshold for Cu (>1000mg/kg dry weight in shoot).

However, the uptake ranges are encouraging when compared to reported toxic concentrations in plants (Vamerali *et al.*, 2010; Liphadzi and Kirkham, 2005). In the case of Pb and Cu, all 28 species were able to withstand the toxic levels. *Croton bonplandianum, Sida longifolia* and *Eclipta alba* accumulated up to 4.5 times the toxicity threshold of Pb and in the case of Cu, *Poa annua* accumulated above 35 times the toxicity threshold. The picture of Cd tolerance across the species evaluated was also good with 4 species crossing the limit in tissues of the shoot and about 8 species doing so in those of the root.

Classification based on Heavy Metal Accumulation Capacity

Depending on the accumulation potential, Baker *et al.*, (2000) categorized plants as excluders (with avoidance mechanisms for element uptake), highly sensitive

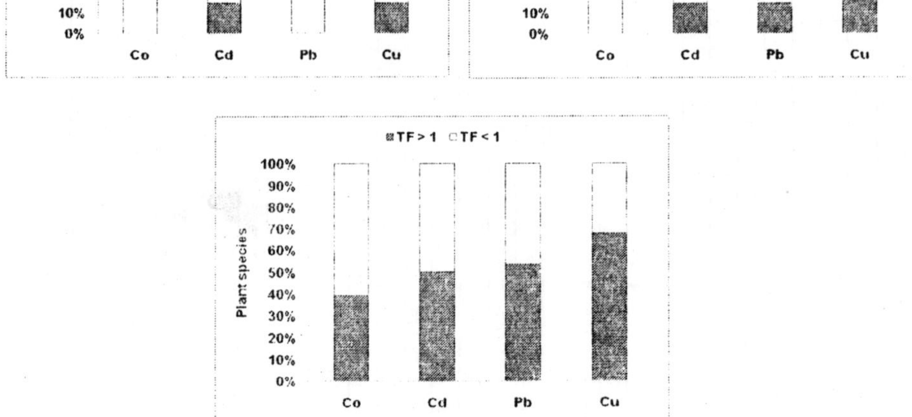

Figure 3: Metal accumulation characteristics of investigated plant species

Table 3
Heavy metal content in plants

Plants	Co		Cd		Pb		Cu	
	Shoot	Root	Shoot	Shoot	Root	Root	Shoot	Root
Amaranthus spinosus	5.61±1.5	5.81±1.42	8.20±1.45	34.06±3.7	94.67±7.3	11.68±1.46	37.03±13.42	86.62±5.5
Argemone Mexicana	6.28±0.46	8.91±0.88	3.98±0.65	31.03±2.1	19.12±0.85	2.47±0.15	31.08±3.06	63.53±7.21
Atriplex lentiformis	10.2±0.34	11.2±3.11	8.11±1.03	33±4.7	85.6±15.3	11.3±4.12	204.6±2.3	112.3±22.5
Boerhaavia diffusa	7.97±1.84	6.63±0.67	8.53±1.2	28.30±4.91	25.30±2.71	6.71±1.24	27.92±3.14	40.32±2.6
Calotropis procera	7.11±3.09	8.62±2.5	6.91±3.9	40.68±9.94	53.64±6.9	12.71±0.67	49.75±17.62	56.96±8.26
Cannabis sativa	8.30±2.33	8.05±4.05	4.49±1.03	28.16±3.6	21.21±2.8	4.48±0.8	30.94±6.35	53.91±4.9
Chenopodium murale	5.08±1.9	5.53±1.09	9.62±1.86	36.56±19.4	13.75±9.6	5.60±0.47	58.84±3.2	37.35±9.06
Crinum asiaticum	8.42±1.2	11.74±4.7	16.3±3.1	127.4±33.2	51.6±16.7	25.4±6.06	73.4±9.43	107.2±27.5
Croton bonplandianum	5.06±0.22	7.04±0.47	3.53±0.7	50.79±25.1	24.11±3.27	4.11±0.25	89.76±7.68	33.55±4.50
Cynodon dactylon	5.24±0.32	3.68±0.17	5.46±0.82	23.00±3.90	16.93±2.87	7.22±0.63	31.53±4.05	26.64±2.43
Datura stramonium	5.24±0.8	5.02±0.6	12.88±4.0	146.15±4.66	40.41±7.02	9.19±2.05	41.33±19.17	44.05±4.78
Eclipta alba	6.22±0.57	11.26±3.42	6.54±0.96	23.05±3.28	55.36±6.4	12.13±1.3	40.73±15.4	97.44±15.6
Euphorbia hirrta	8.42±0.8	11.74±1.3	9.03±0.24	34.18±2.92	17.61±4.05	13.76±0.51	61.55±4.01	46.94±10.70
Gnaphalium luteo-album	6.07±0.73	7.13±0.93	6.04±1.08	27.29±4.3	33.63±9.4	6.14±2.07	25.25±3.21	33.08±3.5
Heliotropium ellipticum	5.40±2.1	4.95±0.6	11.21±5.9	28.82±4.6	58.00±7.14	7.08±2.51	35.16±9.01	53.55±37.15
Jatropha curcas	8.3±1.02	8.05±3.02	3.5±0.1	34.3±1.09	29.1±3.2	1.9±0.5	88.4±21.2	32.6±14.3
Leucas cephalotes	5.18±0.9	11.56±2.01	2.91±0.5	123.53±8.16	50.52±18.31	2.72±0.58	49.11±2.9	53.45±5.02
Lycopersicon esculentum	5.33±1.4	7.29±2.08	5.20±1.21	36.20±2.7	30.49±4.2	4.25±0.9	53.77±8.6	49.98±7.38
Parthenium hysterophorus	4.75±0.75	4.39±0.7	4.27±1.14	30.04±6.04	47.55±39.2	3.98±0.82	36.31±11.2	48.78±31.9
Physalis minima	5.17±0.8	4.82±.81	3.57±0.79	19.82±3.12	18.41±2.4	1.99±0.32	39.73±21.8	26.43±2.32
Poa annua	6.56±1.71	13.58±4.91	9.15±5.01	742.06±32.8	66.27±8.5	16.70±5.52	297.66±14.6	78.09±22.8
Ricinus communis	6.01±1.1	6.51±1.24	3.74±0.67	101.12±75.01	41.68±6.2	3.03±20.64	47.62±20.5	27.44±7.12
Rumex dentatus	5.83±2.25	6.58±1.7	6.68±3.6	34.19±11.6	46.23±9.3	8.03±3.85	45.82±7.27	67.09±21.4
Sida longifolia	4.52±0.7	6.70±0.93	5.47±0.65	92.43±12.11	56.11±6.6	7.54±0.38	75.13±19.31	56.68±6.95
Solanum nigrum	7.59±2.26	6.18±1.4	12.93±1.13	18.07±5.4	9.05±1.57	9.73±3.37	28.49±5.2	23.85±4.05
Tephrosia purpurea	4.14±0.21	3.24±0.41	11.78±2.8	56.49±8.41	22.69±5.6	8.25±2.5	50.73±3.01	26.46±7.1
Tridex procumbens	6.42±0.78	12.11±1.65	7.36±2.78	28.99±6.27	14.49±2.7	13.25±1.12	34.78±9.3	21.48±4.92
Withania somnifera	10.05±4.4	8.96±4.3	9.12±2.4	33.22±9.4	105.69±4.8	10.46±3.57	32.29±11.22	94.74±8.57
Phytotoxicity levels (mg/kg)	60-170[a]		5-10[b]		10-20[b]		15-20[b]	

(a) Adriano 1986; (b) Förstner 1995

indicators (lacking protection mechanisms) and accumulators (with mechanisms of metal tolerance and accumulation capability in above ground biomass). Accordingly excluders, indicators and accumulators should exhibit concentrations of pollutants in shoot biomass that are, respectively, lower, similar and higher than that of the soil. Most of the plants investigated in this study fell under the category of excluders of metals As seen from table 4, all 28 species were excluders of Co; 3 species were indicators of Cu; and 1of Pb. 4 species were accumulators of Cd and Cu; and 3 of Pb. Indicators included *Calotropis procera* for Pb as well as *Chenopodium murale, Jatropha curcas* and *Lycopersicon esculentum* for Cu. *Poa annua* was classified as an accumulator of 2 metals i.e. Pb and Cu. Other accumulators were *Calotropis procera* (Cd, Cu), *Lycopersicon esculentum* (Cd), *Datura stramonium* (Cu) and *Chenopodium murale* (Cd).

Metal Accumulation Characteristics

The following factors are utilized universally to evaluate the phytoextraction/ phytostabilisation potential of the plant species:

1. Bioabsorption Coefficient [BAC] $= \dfrac{\text{Metal content in shoot}}{\text{Metal content in soil}}$

2. Bioconcentration Factor [BCF] $= \dfrac{\text{Metal content in root}}{\text{Metal content in soil}}$

3. Translocation Factor [TF] $= \dfrac{\text{Metal content in shoot}}{\text{Metal content in root}}$

Phytoremediators may be metal specific, performing well for a particular metal or metals in combination. Moreover, phytoremedial potential of plants is also influenced by the mobility and availability of heavy metals in soil and plants, thus, these ratios help to identify the suitability of plants for phytoextraction and phytostabilisation by explaining the accumulation characteristics and translocation properties of metals in plants. Plants exhibiting TF and BAC values greater than 1 are generally considered promising phytoextractors (Fitz and Wenzel, 2002), while plants able to develop extended and abundant root systems and keep the translocation of metals from roots to shoots as low as possible i.e. a high BCF as well as a low TF are potential phytostabilisers (Mendez and Maier, 2008). Yoon *et al.*, (2006) demonstrated *Phyla nodiflora* having TF values of 12 and 6.3 for Cu and Zn as a suitable phytoextractor, and *Gentiana pennelliana* having BCF values of 11, 22 and 2.6 for Pb, Cu and Zn as a phytostabiliser.

The BAC values of the plants studied were above 1 in 7 species for Cu and 4 species each for Cd and Pb. None of the species has a BAC > 1 for Co. On evaluating BCF, 4 species were found to have a ratio greater than 1 in the case of Cd and Cu. All tested plants had low BCF values for Co and Pb indicating that the plants had limited capabilities of mobilizing these metals in the root zone.

TF values indicate that most of the plants tested allocated a greater proportion of metal absorbed to aboveground biomass. 19 species recorded a TF > 1 for Cu; 15 for Pb; 14 for Cd and 11for Co. Several plants were found to have a TF > 1 for multi-metal combinations, indicating the fact that they were naturalized to the contaminated matrix in which they had germinated and grown, and that their metabolism could at least tolerate the multi-metal contamination. The highest TF values recorded were: 11.21 and 3.81 for Cu and Pb, respectively, in *Poa annua*. Similar studies have shown that leaves act as main sinks for heavy metals in hyperaccumulator plants (Psaras and Manetas, 2001; Salt *et al.*,1995). This is attributed to the efficient translocation of heavy metals from roots to shoots and is considered an advantageous strategy as the root system is the primary target in heavy metal toxicity (Baker *et al.*, 1994). Accumulation of potentially toxic metals in the green parts is also thought to be a plant's defensive strategy against herbivores (Liphadzi and Kirkham, 2005).

Plants exhibiting both BAC and TF values > 1 are generally considered promising phytoextractors (Fitz and Wenzel, 2002). *Crinum asiaticum* showed such values for Cd, Pb and Cu; *Chenopodium murale* and *Lycopersicon esculentum* for Cd and Cu; *Poa annua* for Pb and Cu; *Datura stramonium* for Cu; and *Jatropha curcas* for Cd. These species exhibited potential for possible use in phytoextraction. Highest BAC values were observed in *Poa annua* for Cu (5.58) and Pb (1.38) and in *Chenopodium murale* for Cd (2.10). According to Mendez and Maier (2008), a plant suitable for phytostabilisation should have BCF > 1 and TF < 1. In plants with low TF, there is less translocation of metals to the aboveground portions. The lower translocation of metals to the above-ground portions may be due to immobilization of metals in roots by vacuole sequestration or cell wall binding, thereby preventing interaction with high-molecular-weight compounds in the plant cell cytoplasm (Salt *et al.*, 1995). High BCF and low TF values were observed in *Calotropis procera* for Cd and Cu; *Poa annua* for Cd; and in *Withania somnifera* for Cu, thus, establishing them as good phytostabilisers for these metals and metal combinations.

Chlorophyll and Free Proline Patterns in Leaf Samples

When present in high concentration in soil, metals like Cd, Pb, Zn, Cr, etc. show potential toxic effects on overall growth and metabolism of plants (Agrawal and Sharma, 2006). Chlorophyll, essential for photosynthesis, is directly influenced by environmental factors. Decreased chlorophyll content in several plant species under the impact of heavy metals is reported (Oncel *et al.*, 2000). Nearly all the species tested had lower total chlorophyll content than control plants (Figure 5). Heavy metals such as Cu, Co and Zn are essential for normal plant growth and development since they are constituents of many enzymes and other proteins. However, elevated concentrations of both essential and non-essential heavy metals in the soil can lead to toxicity symptoms and growth inhibition in most plants (Hall, 2002). Phytotoxicity is mainly associated with non-essential metals like As, Cd, Cr and Pb, which generally have very low toxicity thresholds (Clemens, 2006) and lower threshold values for hyperaccumulation (especially for Cd) than the other metals.

Heavy metals are known to interfere with chlorophyll synthesis either through direct inhibition of an enzymatic step or by inducing deficiency of an essential nutrient (van Assche and Clíjsters, 1990). The amount of chlorophyll was reduced in *Triticum aestivum* on Cu-enriched soil (Lanaras *et al.*, 1993), and in *Brassica oleracea* var. Botrytis exposed to Cu, Co and Cr (Chatterjee and Chatterjee, 2000). Reduction in total chlorophyll levels in plants due to Pb, Cu, Cd and Hg has been observed (Zengin and Munzuroglu, 2005; Faheed, 2005). Pang et al. (2003) showed that high proportions of lead and zinc (Pb/Zn) tailing greatly inhibited the leaf growth and photosynthesis of Vetiver grass but stimulated proline levels.

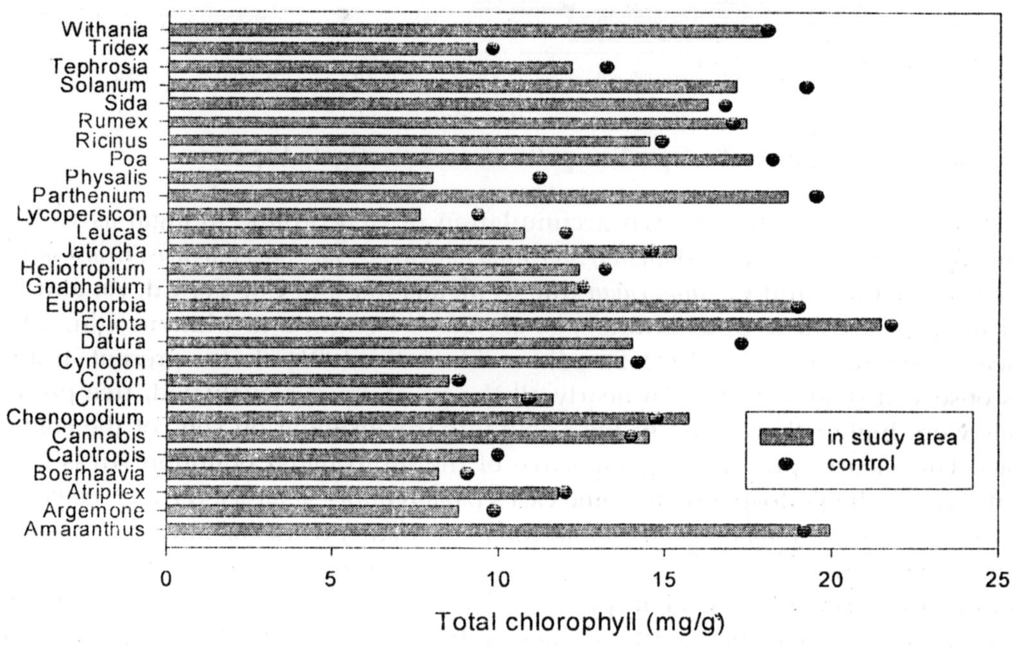

Figure 5. Chlorophyll patterns in investigated species

Accumulation of free proline in response to heavy metal exposure seems widespread among plants. Proline accumulation, accepted as an indicator of environmental stress, is also considered to have important protective roles. The exact mechanism of how proline accumulation helps the plant to cope up with heavy metal stress is difficult to elucidate. However, available evidences suggest Proline increases the stress tolerance of plants through such mechanisms as osmoregulation, protection of enzymes against denaturation, and stabilization of protein synthesis (Kavi *et al.*, 2005; Ashraf and Foolad, 2007).

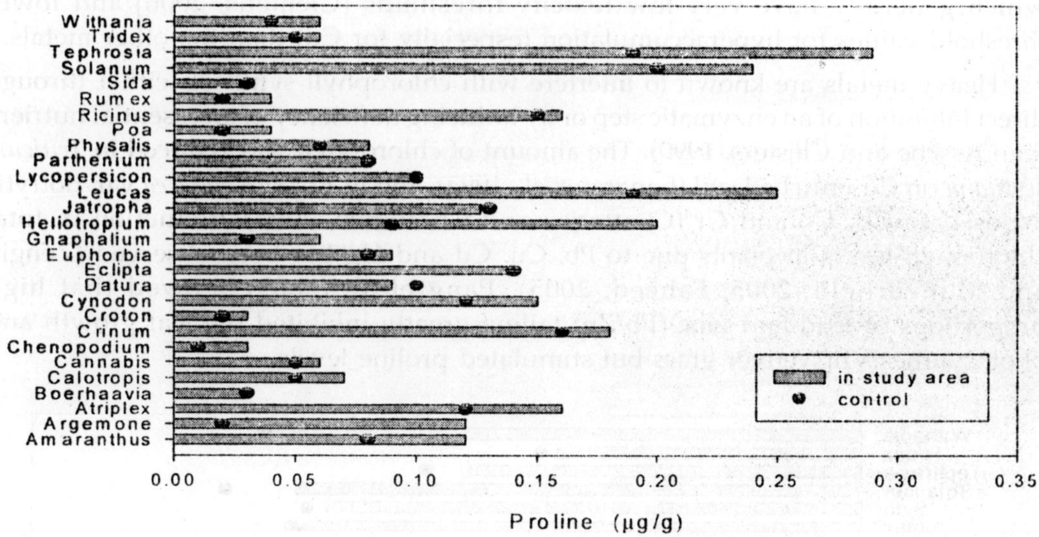

Figure 6. Free proline patterns in investigated species

Detailed studies have shown accumulation of free proline in algal as well as plant species *viz.* — *Crinum asiaticum, Spirulina platensis, Phaseolus vulgaris, Eruca sativa, Helianthus annuus,* and *Cajanus cajan* due to metals like Pb, Cu, Co, Cd, Zn and Hg (Varun *et al.*, 2010a & b; Choudhary *et al.*, 2006; Zengin and Munzuroglu, 2005; Faheed, 2005; Kastori *et al.*, 1992; Saradhi and Saradhi, 1991). In the present study it was observed (Figure 6) that in nearly all the plant species evaluated, free proline levels were higher than those in their counterparts growing on soils of the control zones. This clearly puts us in perspective of the multi-metal pollution stress these plants face in the pedosphere. In some cases (*Amaranthus spinosus, Ricinus communis, Argemone mexicana, Euphorbia hirrta* and *Rumex dentatus*) it was as much as twice the control levels. Similar results were seen in response to Cu, Cd and Zn in metal-tolerant and non-tolerant ecotypes of *Silene vulgaris*; the constitutive proline concentration in leaves being 5 to 6 times higher in the metal-tolerant ecotype than in the non-tolerant ecotype (Schat *et al.*, 1997).

It has been observed that at a certain interval of contamination, plants are generally able to defend themselves from heavy metals. According to Handique and Handique (2009), toxic heavy metals like Pb. Hg and Cd induced higher proline accumulation in lemongrass following short term exposure (two months after transplantation) than long term exposure (nine months after transplantation). This can be a possible explanation for the fact that in about 70 % of the plants studied the difference in proline levels was not very marked compared to the control plants. In all probability these plants being annuals and perennials had adapted themselves to the higher levels of heavy metals in the soil. On the other hand, markedly higher proline levels were obtained in plants with a shorter life span i.e. those that had grown later than the former *viz.* — *Heliotropium ellipticum, Chenopodium murale, Tephrosia purpurea, Solanum nigrum* and *Leucas cephalotes* which is also in concurrence with the results of the aforementioned study.

PROSPECTS FOR PHYTOREMEDIATION

Phytoremediation has gained acceptance as a technology and has been acknowledged as an important area of research in the past fifteen years. New knowledge and plant material obtained from research is already being implemented for phytoremediation in the field. Nonetheless, its basic processes are still largely not clear and hence further research is required to improve its field performance. Information collected from basic research at physiological, biochemical and genetic level in plants will be helpful in understanding the processes of passive adsorption, active uptake, translocation, accumulation and chelation mechanisms. Research aimed at better understanding of the interactive roles among plants roots and microbes will help scientists to utilize their integrative capacity for soil decontamination. The role of agronomic practices like tilling, fertililizing etc. in improving phytoremedial capabilities of plants needs to be explored. Application of biosurfactants and metal chelators that can alter the bioavailable fraction of metals in the rhizosphere has been documented and is a promising avenue to be explored. Further, genetic

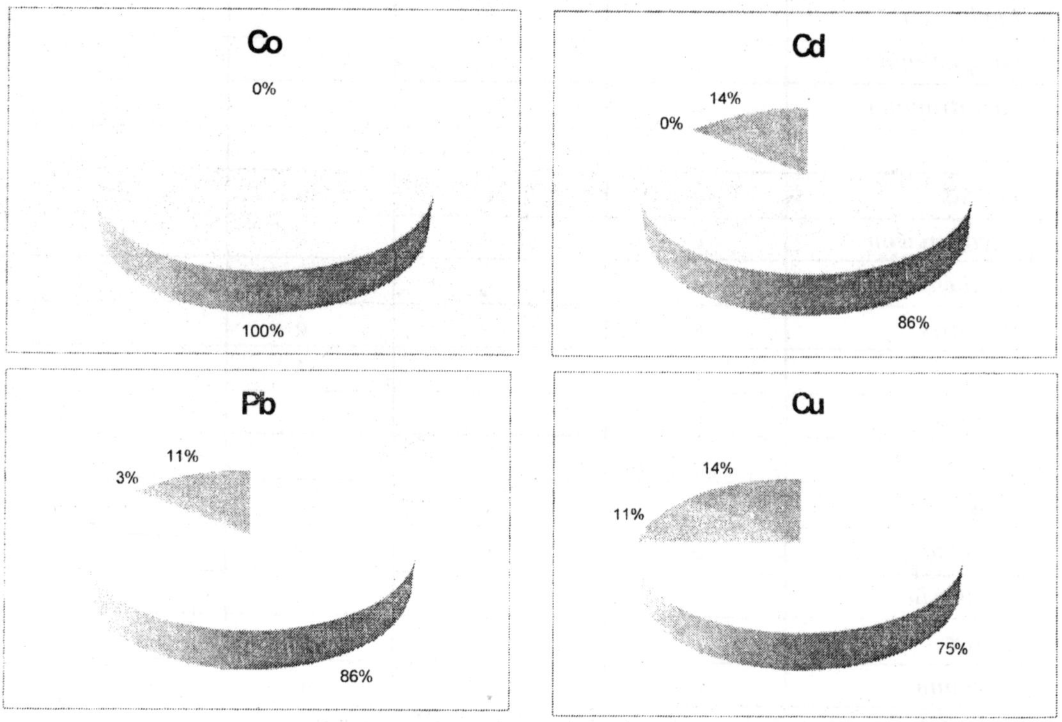

Figure 7. Heavy metal accumulation capacity of investigated species.

Table 4

Classification of investigated species based on their Heavy metal accumulation capacity

Plant species	Co	Cd	Pb	Cu
Amaranthus	¢	¢	¢	¢
Argemone	¢	¢	¢	¢
Atriplex	¢	¢	¢	¢
Boerhaavia	¢	¢	¢	¢
Calotropis	¢	A	I	A
Cannabis	¢	¢	¢	¢
Chenopodium	¢	A	¢	I
Crinum	¢	A	A	A
Croton	¢	¢	¢	¢
Cynodon	¢	¢	¢	¢
Datura	¢	¢	¢	A
Eclipta	¢	¢	¢	¢
Euphorbia	¢	¢	¢	¢
Gnaphalium	¢	¢	¢	¢
Heliotropium	¢	¢	¢	¢
Jatropha	¢	¢	A	I
Leucas	¢	¢	¢	¢
Lycopersicon	¢	A	¢	I
Parthenium	¢	¢	¢	¢
Physalis	¢	¢	¢	¢
Poa	¢	¢	A	A
Ricinus	¢	¢	¢	¢
Rumex	¢	¢	¢	¢
Sida	¢	¢	¢	¢
Solanum	¢	¢	¢	¢
Tephrosia	¢	¢	¢	¢
Tridex	¢	¢	¢	¢
Withania	¢	¢	¢	¢

'¢' Excluder (BAC d″ 0.90); 'I' Indicator (0.91 d″ BAC d″ 1.10); 'A'- Accumulator (BAC e″ 1.10) evaluation of hyperaccumulators growing in metal contaminated soil and associated microbes would provide the researchers with a gene pool to be used in genetic manipulation of other non accumulators and production of transgenics. As more results demonstrating the effectiveness of phytoremediation become available, its use may continue to grow, reducing clean up costs and enabling the clean-up of more sites with the limited funds available.

SUMMARY

The results of this study confirm that the ill-monitored activities of the industry have caused significant accumulation of heavy metals in both soils and native wild flora suggesting that industrial pollution is an important source contributing to urban environmental deterioration in the study area and should be a matter of great concern. Proper monitoring is urgently needed. The results of the present investigation indicate that individual elements displayed remarkably different patterns of accumulation in soils. Native species have adapted themselves to multi-metal contamination in the pedosphere and have also evolved strategies to protect their metabolic integrity as shown by chlorophyll and free proline patterns and to avoid phytotoxicity as they were able to tolerate and accumulate metal concentrations far above the reported phytotoxicity levels especially in the case of Cd, Pb, and Cu. Seasonal species showed higher proline accumulation than perennials. Uptake as well as translocation patterns of metals in plants differed remarkably. Plants had difficulties in mobilizing Co and Pb in the root zone as seen from their low BCF values (root to soil ratio of metal). Most of the plants fall under the category of excluders of metals. The following species can be recommended as phytoextractors — *Crinum asiaticum* for Cd, Pb and Cu; *Chenopodium murale* and *Lycopersicon esculentum* for Cd and Cu; *Poa annua* for Pb and Cu; *Datura stramonium* for Cu; and *Jatropha curcas* for Cd either singly or for these multi-metal combinations. Furthermore, *Poa annua* exhibited the greatest phytoextraction potential for Cu and *Chenopodium murale* for Cd among the species investigated. The following species can be recommended as phytostabilisers — *Calotropis procera* for Cd and Cu; *Poa annua* for Cd and *Withania somnifera* for Cu. None of the species assessed showed potential for phytoextraction/phytostabilisation of Co. Similarly, no promising phytostabilisers were observed for Pb in the native flora studied.

ACKNOWLEDGEMENT

Financial support from University Grants Commission [F. no. 35-47/2008(SR)] is duly acknowledged. The authors would also like to thank the Principal, St. John's College and the Head, Department of Botany for providing facilities.

REFERENCES

1. Adriano, D.C. (1986). Trace elements in the terrestrial environment. Springer, New York.
2. Agrawal, V., Sharma, K. (2006). Phytotoxic effects of Cu, Zn, Cd and Pb on in vitro regeneration and concomitant protein changes in *Holarrhena antidysentrica*. Biol. Plant. 50, 307-310.
3. Ashraf, M., Foolad, M.R. (2007). Roles of glycine, betaine and proline in improving plant abiotic stress resistance. Environ. Exp. Bot. 59, 206-216.
4. Baker, A.J.M., Brooks, R.R. (1989). Terrestrial higher plants which hyperaccumulate metallic elements—A review of their distribution,ecology and phytochemistry. Biorecovery 1:81–126.
5. Baker, A.J.M., McGrath, S.P., Reeves, R.D., Smith, J.A.C. (2000). Metal hyperaccumulator plants: A review of the ecology and physiology of a biological resource for phytoremediation of metal-polluted soils, in Terry, N., Banuelos, G. (Eds.), Phytoremediation of Contaminated Soil and Water, Lewis Publishers, CRC, Boca Raton, pp. 85–107.
6. Baker, A.J.M., Reeves, R.D., Hajar, A.S.M. (1994). Heavy metal accumulation and tolerance in British populations of the metallophyte *Thlaspi caerulescens* J. & C. Presl (Brassicaceae). New Phytol. 127, 61-68.
7. BIO-WISE, (2000). Contaminated land remediation : A review of biological technology; (London Dept. of Trade and Industry, Govt. of UK).
8. Brooks, R.R., Baker, A.J.M., Malaisse, F. (1992). The unique flora of the Copper Hills of Zaire. Res. Explor. 8, 338-351.
9. Chatterjee, J., Chatterjee, C. (2000). Phototoxicity of cobalt, chromium and copper in cauliflower. Environ. Pollut. 109, 69–74.
10. Chirenje, T., Ma, L.Q., Reeves, M., and Szulczewski, M. (2004). Lead distribution in near surface soils of two Florida cities: Gainesville and Miami. Geoderma. 119, 113–120.
11. Choudhary, M., Jetley, U.K., Khan, M.A., Zutshi, S., Fatma, T. (2006). Effect of heavy metal stress on proline, malondialdehyde, and superoxide dismutase activity in the cyanobacterium Spirulina platensis-S5. Ecotoxicol. Environ. Saf. 66, 204-209.
12. Clemens, S. (2006). Toxic metal accumulation, responses to exposure and mechanisms of tolerance in plants. Biochimie. 88, 1707–1719.
13. Crnkoviæ, D., Ristiæ, M., Antonoviæ, D. (2006). Distribution of Heavy Metals and Arsenic in Soils of Belgrade (Serbia and Montenegro). Soil Sedimen. Contam. 15, 581–589.
14. Cunningham, S.D. and Ow, D.W. (1996). Promises and Prospects of Phytoremediation-Update on Biotechnology. Plant Physiology110:715-719.
15. D'Souza, R., Varun, M., Masih, J., Paul, M.S. (2010). Identification of *Calotropis procera* L. as a potential phytoaccumulator of heavy metals from contaminated soils in Urban North Central India. J Haz. Mat. DOI: 10.1016/j.jhazmat.2010.08.056
16. Datta, R.and Sarkar, D. (2004). Biotechnology in Phytoremediation of Metal-contaminated soils; Proc. Indian natn. Sci. Acad. B70 .(1).p.99-108.
17. Del Castilho, P., Chardon, W.J. and Salomons, W. (1993). Influence of cattle-manure slurry application on the solubility of cadmium, copper, and zinc in a manured acidic, loamy-sand soil. J. Environ. Qual. 22, 689-697.

18. Desouza, M.P.; Huang, C.P.; Chee, N. and Terry, N. (1999). Rhizosphere bacteria enhance the accumulation of Selenium and Mercury in wetland plants. Planta, , vol. 209, no. 2, p. 259-263.
19. Dietz, A.C., Schnoor, J.L. (2001). Advances in phytoremediation. Environ. Heal. Perspect. 109,163–168.
20. EPA (Environmental Protection Agency), (1997). Electro kinetic laboratory and field processes applicable to radioactive and hazardous mixed waste in soil and groundwater. EPA 402/R-97/006. Washington, DC.
21. Erakhrumen; Agbontalor, A. (2007). Phytoremediation: an environmentally sound technology for pollution prevention, control nd remediation in developing countries. Educ. Res. Rev. 2(7): 151-156.
22. Faheed, F.A. (2005). Effect of lead stress on the growth and metabolism of *Eruca sativa* M. Seedlings. Acta Agronomica Hungarica. 53 (3), 319-327.
23. Ferner, D.J. (2001). Toxicity, heavy metals. eMed. J. 2(5): 1.
24. Fitz,W.J., Wenzel, W.W. (2002). Arsenic transformation in the soil–rhizosphere–plant system, fundamentals and potential application of phytoremediation, J. Biotechnol. 99, 259–278.
25. Förstner, U. (1995). Land contamination by metals: global scope and magnitude of problem, in: Allen, H.E., Huang, C.P., Bailey, G.W., Bowers, A.R. (Eds.), Metal speciation and contamination
26. Halim, M., Conte, P., Piccolo, A. (2003). Potential availability of heavy metals to phytoextraction from contaminated soils induced by exogenous humic substances. Chemosphere 52, 265–275.
27. Hall, J.L. (2002). Cellular mechanisms for heavy metal detoxification and tolerance. J. Exp. Bot. 53, 1–11.
28. Hamlin, R. (2002). Phytoremediation literature review. Pollution and environmental awareness in The United States.
29. Handique, G.K., Handique, A.K. (2009). Proline accumulation in lemongrass (*Cymbopogon flexuosus* Stapf.) due to heavy metal stress. J. Environ. Biol. 30(2), 299-302.
30. Hartley, W., Dickinson, N.M., Clemente, R., French, C., Pierce, T.G., Sparke, S., Lepp, N.W. (2009). Arsenic stability and mobilization in soil at an amenity grassland overlying chemical waste (St. Helens, UK). Environ. Pollut. 157, 847–856.
31. Henry, H. (2006). Natural Revegetation of an Aged Petroleum Land farm Impacted with PAHs and Heavy Metals: Ecological restoration, Remediation, and Risk, PhD Thesis, University of Cincinnati.
32. Hinchman, R.R., Negri, M.C., Gatliff, E.E. (1998). Phytoremediation : using green plants to clean up contaminated soil, ground water and wastewater. Phytoremediation. Argonne National Laboratory. 16: (1) 1-10.
33. Imperato, M., Adamo, P., Naimo, D., Arienzo, M., Stanzione, D., Violante, P. (2003). Spatial distribution of heavy metals in urban soils of Naples city (Italy). Environ. Pollut. 124, 247–256.
34. Institute of Environmental Conservation and Research INECAR (2000). Position Paper Against Mining in Rapu-Rapu, Published by INECAR, Ateneo de Naga University, Philippines (www.adnu.edu.ph/Institutes/Inecar/pospaper1.asp)

35. Kartori, R., Petroviæ, M., Petroviæ N. (1992). Effect of excess lead, cadmium, copper, and zinc on water relations in sunflower *J. Plant Nutr.* 15(11), 2427 - 2439.
36. Kavi, K.P.B., Sangam, S., Amrutha, R.N., Laxmi, P.S., Naidu, K.R., Rao, K.R.S.S., Rao, S., Reddy, K.J., Theriappan, P., Sreenivasulu, N. (2005). Regulation of proline biosynthesis, degradation, uptake and transport in higher plants: Its implications in plant growth and abiotic stress tolerance. Curr. Sci. 88, 424-438.
37. Lanaras, T., Moustakas, M., Symeonl'dl's, L., Diomantoglou, S., Karatagll's, S. (1993). Plant metal content, growth responses and some photosynthetic measurements of field-cultivated wheat growing on ore bodies enriched in Cu. Physiol. Plant. 88, 307–314.
38. Lasat, M.M. (2002). Phytoremediation of toxic metals: a review of biological mechanisms. J Environ. Qual. 31,109–120.
39. Lenntech Water Treatment and Air Purification. Water Treatment 2004 published by Lenntech, Rotterdamseweg, Netherlands, (www.excelwater.com/thp/filters/Water-Purification.htm).
40. Liphadzi, M.S., Kirkham, M.B. (2005). Phytoremediation of soil contaminated with heavy metals: a technology for rehabilitation of the environment, S. Afr. J. Bot. 71(1), 24-37.
41. Lombi, E., Zhao, F.J., Dunham, S.J., McGrath, S.P. (2001). Phytoremediation of heavy metal contaminated soil: natural hyperaccumulation versus chemically enhanced phytoextraction, J. Environ. Qual. 30, 1919-1926.
42. Madrid, L., Dìaz-Barrientos, E., Madrid, F. (2004). Metals in urban soils of Sevilla: seasonal changes and relations with other soil components and plant contents. Eur. J. Soil Sci. 55, 209–217.
43. McCluggage, D. (1991). Heavy Metal Poisoning, NCS Magazine, Published by The Bird Hospital, CO, U.S.A. (www.cockatiels.org/articles/Diseases/metals.html).
44. McGrath, S.P., Zhao, F.J. (2003). Phytoextraction of metals and metalloids from contaminated soils. Curr. Opin. Biotech. 14, 277–282.
45. McLaughlin, M.J., Hamon, R.E., McLaren, R.G., Speir, T.W., Rogers, S.L. (2000). Review: A bioavailability based rationale for controlling metal and metalloid contamination of agricultural in Australia and New Zealand. Aust. J. Soil Res. 38, 1037-1086.
46. Mellem, J.J., Baijnath, H., Odhav, B. (2009). Translocation and accumulation of Cr, Hg, As, Pb, Cu and Ni by Amaranthus dubius (Amaranthaceae) from contaminated sites. J. Environ. Sci. Heal. 44, 568–575.
47. Mendez, M.O., Maier, R.M. (2008). Phytostabilization of mine tailings in arid and semiarid environments—an emerging remediation technology. Environ. Heal. Perspect. 116, 278–283
48. Nedunuri, K.V., Lowell, C., Meade, W., Vonderheide, A.P., Shann, J.R. (2010). Management practices and Phytoremediation by Native grasses. Int. J. Phytorem. 12, 200–214.
49. Nolan, K. (2003). Copper Toxicity Syndrome, J. Orthomol. Psychiatry 12(4): 270 – 282.
50. Nouri, J., Khorasani, N., Lorestani, B., Karami, M., Hassani, A. H., Yousefi, N. (2009). Accumulation of heavy metals in soil and uptake by plant species with phytoremediation potential. Environ. Earth Sci. 59, 315–323.
51. Oncel, I., Keles, Y., Ustun, A.S. (2000). Interactive effects of temperature and heavy metal stress on the growth and some biochemical compounds in wheat seedlings. Environ. Pollut. 107, 315–320.

52. Ouzounidou, G. (1994). Root growth and pigment composition in relationship to element uptake in Silene compacta plants treated with copper. J. Plant Nutr. 17, 933-943.

53. Pang, J., Chan, G.S.Y., Zhang, J., Liang, J., Wong, M. H. (2003). Physiological aspects of vetiver grass for rehabilitation in abandoned metalliferous mine wastes. Chemosphere. 52(9), 1559-1570.

54. Porêbska, G., Ostrowska, A. (1999). Heavy metal accumulation in wild plants: implications for phytoremediation. Pol. J. of Environ. Stud. 8(6), 433-442.

55. Prasad, M.N.V., De Oliveira-Freitas, H.M. (2003). Metal hyperaccumulation in plants—Biodiversity prospecting for phytoremediation technology. Electr J Biotech 6:285–321

56. Prasad, M.N.V. (2004). Phytoremediation of metals and radionuclides in the environment; the case for natural hyperaccumulators, metal transporters, soil amending chelators and transgenic plants. In: Heavy metal stress in plants: from biomolecules to ecosystems. Heidelberg, Springer-Verlag, 2nd ed., p.345-392.

57. Psaras, G.K., Manetas, Y. (2001). Nickel localization in seeds of the metal hyperaccumulator *Thlaspi pindicum* Hausskn. Ann. Bot., 88, 513-516.

58. Pulford, I.D., Watson, C. (2003). Phytoremediation of heavy metalcontaminated land by trees—a review. Environ. Int. 29, 529–540.

59. Raskin, I., Smith, R.D., Salt, D.E. (1997). Phytoremediation of metals: Using plants to remove pollutants from the environment. Curr. Opin. Biotechnol. 8(2), 221-226.

60. Saifullah, M.E., Qadir, M., de Caritat, P., Tack, F.M.G., Du Laing, G., Zia, M.H. (2009). EDTA-assisted Pb phytoextraction. Chemosphere 74, 1279–1291.

61. Salt, D.E. and Kramer, U. (2000). Mechanisms of metal hyperaccumulation in plants. In: Raskin I, Ensley B (eds) Phytoremediation of toxic metals: using plants to clean up the environment. Wiley, New York, pp 231–246.

62. Salt, D.E., Blaylock, M., Kumar, N.P.B.A., Viatcheslav, D., Ensley, B.D. (1995). Phytoremediation a novel strategy for the removal of toxic metals from the environment using plants. Biotechnol. 13, 468-474.

63. Samarghandi, M.R., Nouri, J., Mesdaghinia, A.R., Mahvi, A.H., Nasseri, S., Vaezi, F. (2007). Efficiency removal of phenol, lead and cadmium by means of $UV/TiO_2/H_2O_2$ processes. Int. J. Environ. Sci. Technol. 4(1), 19–25

64. Saradhi, A., Saradhi, P.P. (1991). Proline accumulation under heavy metal stress. J. Plant Physiol. 138, 554-558.

65. Saraswat, S., Rai, J.P.N. (2009). Phytoextraction potential of six plant species grown in multimetal contaminated soil. Chem. Ecol. 25(1), 1–11.

66. Sauve, S., McBride, M.B., Norvell, W.A. and Hendershot, W.H. (1997). Copper solubility and speciation of in situ contaminated soils: effects of copper level, pH and organic matter. Water, Air Soil Pollut. 100, 133-149.

67. Schat, H., Sharma, S.S., Vooijs, R. (1997). Heavy metal-induced accumulation of free proline in a metal-tolerant and non-tolerant ecotype of *Silene vulgaris*. Physiol. Plant. 101, 477–482.

68. Srivastava, P.C., Gupta, U.C. (1996). Trace Elements in Crop Production. Science Publishers Inc., Lebanon, New Hampshire, USA.

69. Tijhuis, L., Brattli, B., Saether, O.M. (2002). A geochemical survey of topsoil in the city of Oslo, Norway. Environ. Geochem. Heal. 24, 67–94.

70. Turner, A.P. (1994). The responses of plants to heavy metals. In: Ross, S.M (Ed.) 'Toxic Metals in Soil-Plant Systems'. John Wiley and Sons, Chichester. pp. 153-187.
71. United States Department of Labor 2004. Occupational Safety and Health Administration (OSHA); Safety and Health Topics: Heavy metals, USDOL Publication, Washington, D.C. (www.osha.gov/SLTC/metalsheavy/index.html).
72. Uraguchi, S., Watanabe, I., Yoshitomi, A., Kiyono, M., Kuno, K. (2006). Characteristics of cadmium accumulation and tolerance in novel Cd-accumulating crops, *Avena strigosa* and *Crotalaria juncea*. J. Exp. Bot. 57(12), 2955-2965.
73. US Environmental Protection Agency 2002. Standards for the use or disposal of sewage sludge. Publication 50313. http://www.gpoaccess.gov/cfr/index.html
74. Vamerali, T., Bandiera, M., Coletto, L., Zanetti, F., Dickinson, N.M., Mosca, G. (2009). Phytoremediation trials on metal- and arsenic-contaminated pyrite wastes (Torviscosa, Italy). Environ. Pollut. 157, 887–894.
75. Vamerali, T., Bandiera, M., Mosca, G. (2010). Field Crops for phytoremediation of metal-contaminated land — A review. Environ. Chem. Lett. 8, 1–17.
76. Van Assche, F., Clijsters, H. (1990). Effect of metals on enzyme activity in plants. Plant Cell Environ. 13, 195–206.
77. Varun, M., D'Souza, R., Kumar, D., Paul, M.S. (2010a). Bioassay as monitoring system for lead phytoremediation through *Crinum asiaticum* L. Environ. Monit. Assess. DOI: 10.1007/s10661-010-1696-9
78. Varun, M., D'Souza, R., Kumar, D., Paul, M.S. (2010b). Effect of Lead (Pb) stress on physio-biochemical processing of *Crinum asiaticum*. J. Holis. Environ. (accepted)
79. Viard, B., Pihan, F., Promeyrat, S., Pihan, J.C. (2004). Integrated assessment of heavy metal (Pb, Zn, Cd) highway pollution: bioaccumulation in soil, Graminaceae and land snails. Chemosphere. 55, 1349–1359.
80. Yang, B., Shu, W., Ye, Z., Lan, C., Wong, M. (2003). Growth and Metal Accumulation in Vetiver and two Sebania species on Lead/Zinc Mine Tailings, Chemosphere. 52, 1593-1600.
81. Yoon, J., Cao, X., Zhou, Q., Ma, L.Q. (2006). Accumulation of Pb, Cu, and Zn in native plants growing on a contaminated Florida site. Sci. Total Envir. 368, 456–464.
82. Zengin, F.K., Munzuroglu, O. (2005). Effects of some heavy metals on content of Chlorophyll, Proline and some antioxidant chemicals in bean (Phaseolus vulgaris L.) seedlings. Acta Biologica Cracoviensia Series Botanica 47/2: 157–164.

9

A Novel Route for Waste Water Treatment:
Photo–Assisted Fenton Degradation of a Reactive Dye

Menka Surana and B.V. Kabra

INTRODUCTION

Today, protection of environment has become a major and crucial factor. The world is facing the challenge of purification of water and other resources. A lot of industries especially textile, dyeing and printing industries have generated environmental problems like aesthetic pollution, eutrophication and perturbation of aquatic ecosystem. Dyes make our world beautiful but they bring pollution. Synthetic dyes are present in many spheres of our everyday life and their applications are continuously growing, e.g., in various branches of textile industry, leather tanning industry, in paper production, in food technology, in agricultural research, in photo-electrochemical cells, in hair colorings etc. Waste waters originating from dyes production and application industries pose a major threat to the surrounding ecosystems because of their toxicity and potentially carcinogenic nature (Kusic *et al.*, 2005). Dye molecules undergo chemical changes as well as biological changes in the aquatic system, consume dissolved oxygen and thus disturb the eco-system (Solmaz *et al.*, 2006). The discharge of highly colored waste water is not only aesthetically displeasing, but it also impedes light penetration thus upsetting biological processes within a stream. Furthermore their presence in drinking water constitutes a potential human health hazard.

DYE COMPOUNDS

Dyes are exclusively organic compounds. Foundation of synthetic dye is attributed to William Henry Perkin on his discovery of a purple dye named Aniline purple, which was later known as Mauveine in 1956.

A good dye gives a color that is fast, which means it resists fading by water, light, perspiration and other factors. Main dye classes according to their applications are categorized as acid dyes, azoic dyes, basic dyes, direct dyes, reactive dyes, disperse dyes, mordant dyes, sulphur dyes and vat dyes.

The most typical dyestuff is reactive dyes, which are estimated to have the highest worldwide market share (Zollinger, 2003).

Reactive dyes are extensively used in the textile industry, fundamentally due to the capacity of their reactive groups to bind on textile fibers by covalent bonds. The dyes are mainly used for dyeing cellulosic fibers, such as cotton and rayon, but are also used for dyeing silk, wool, nylon and leather (Yang *et al.*, 2005). It was estimated that approximately 50% of unbound reactive dyes would remain in the waste water after dyeing process (He *et al.*, 2007).

SYNTHETIC TEXTILE WASTEWATER

Synthetic dye solutions had been used by most researchers in their investigation of treatment technologies since synthetic solutions was useful in obtaining information on how individual dyes react to different types of treatment. Apart from this, constant composition of a synthetic solution enables the specific study of treatment efficiency on a particular treatment technology.

TREATMENT METHODS

Most of the pollutants, except color, can be reduced by chemical, physical or biological methods. Unfortunately, water soluble dyes (eg. Reactive dyes) are not easily removed in conventional biological treatment processes or the conventional physiocoagulation methods due to their high solubility and non-biodegradability (Juang *et al.*, 1997). Similarly, application of chemical treatment is limited due to their high operating cost.

ADVANCED OXIDATION PROCESSES

The enormous diversity of toxic and organic pollutants of various chemical compositions eliminates the possibility of using a universal treatment method and has led to the fast development of special treatment methods for decontamination (Ming-Ho, 2005).

Advanced oxidation processes (AOPs) have provided innovative, cost-effective catalyzed chemical oxidation for treating pollutants in low or high concentration from contaminated soil, sludge and water. AOPs are based on the generation of hydroxyl radical (·OH) which has a high oxidation potential (2.8 V) that completely convert organic contaminant into CO_2, H_2O and inorganic ions or biodegradable compounds.

Common AOPs which are widely used are UV irradiation, H_2O_2/ UV (Ugurlu and Kula, 2007), TiO_2/UV (Chatterjee and Dasgupta, 2005), Ozonation (Haperman *et al*, 1997), Fenton reagent (Kavitha and Palanivelu, 2005) or UV- Fenton (Gabriel and Hong, 2008) process etc.

Fenton Reagent has found to be a promising treatment method for the effective decolourization and degradation of dyes. Fenton reagent exhibits a number of features that make its use advantageous as compared to other methods such as high degree of catalytic efficiency, high degree of specificity and absence of side reactions. Compared with other oxidation processes, the Fenton Reagent is relatively cheap, easily operated and maintained (Swaminathan *et al.*, 2003). It can be used to degrade many pesticides (Segura *et al.*, 2008), harmful chemicals (Bag *et al.*, 2009) and dyes (Orozcoa *et al.*, 2008). It has the advantages of coagulation and catalytic oxidation, as well as being able to generate oxygen in water. Fenton and Fenton type processes have proven to yield very good results either for complete mineralization of azo dyes or for their transformation into less complex structures that are more easily biodegradable.

EXPERIMENTAL

Materials

One of the most common dye which is widely used in textile industries is Reactive Orange-16 (RO16). RO16 bears an azo group as chromophore and a sulphato- ethyl sulphone as the reactive group. This dye was highly recommended for printing process and usually used for coloring silk and cotton. As, it causes great potential of pollution of water environment, a proper treatment is necessary before discharge into the environment. Therefore, it is planned to undertake the title investigation.

For the present studies the commercial disazo dye Reactive Orange-16 (C.I No. 17757, Aldrich) (Fig. 1) having 95% dye content was used. Photocatalyst $FeSO_4$ (Merck, 99% purity) and H_2O_2 (Merck, 6% purity) were used for photocatalytic degradation. For the photobleaching process, 1×10^{-3}M stock solution of dye Reactive Orange-16 was prepared in double distilled water and diluted as required. The desired pH of the solution was adjusted by the addition of previously standardized sulphuric acid and sodium hydroxide solutions. All laboratory reagents were of analytical grade.

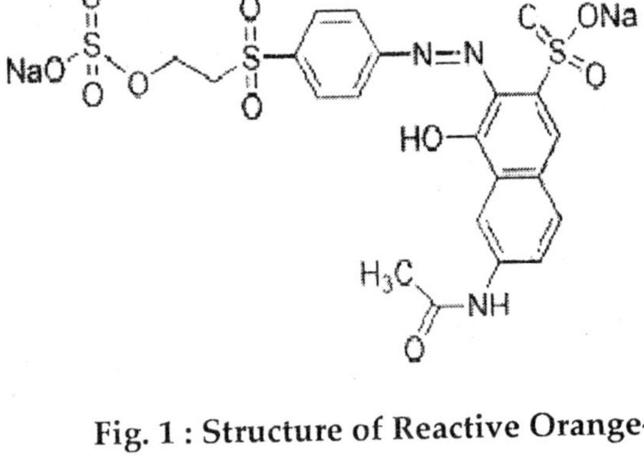

Fig. 1 : Structure of Reactive Orange-16

Molecular formula – $C_{20}H_{17}N_3Na_2S_3O_{11}$ Molecular weight = 617.54

Procedure and Analysis

The reaction mixture was prepared by taking 3 ml of Reactive Orange-16 dye solution (1×10^{-3} M), 3 ml of $FeSO_4$ (1×10^{-3} M), 0.5 ml of H_2O_2 (30 %) in a round bottom flask. The total volume of the reaction mixture was made 100 ml by adding double distilled water. The concentration of different ingredients in the reaction mixture was (Dye) $= 3 \times 10^{-5}$ M, ($FeSO_4$) $= 3 \times 10^{-5}$ M, (H_2O_2) $= 4.89 \times 10^{-2}$ M.

To carry out the photobleaching, the reaction mixture was irradiated under light source (2 x 200 W Tungsten lamps). Water filters were used to cut off thermal radiation. The pH of the solution was measured by pH meter (Systronics, 106). The progress of the reaction was observed at definite time intervals by measuring absorbance using spectrophotometer (Schimadzu, UV- 1700, Pharmaspec) at 490 nm. The rate of decrease of colour with time was continuously monitored. After complete mineralization, the presence of NO_2^-, NO_3^-, SO_4^{2-} ions and evolution of CO_2 were tested by standard procedure.

RESULTS AND DISCUSSION

Control experiments (in absence of photocatalyst, light) confirm the necessity of photocatalyst, light to follow the photocatalytic path for the photobleaching of dye.

The photocatalytic degradation of Reactive Orange-16 was observed at 490 nm. The optimum conditions for the photobleaching of dye were (Dye) $= 3 \times 10^{-5}$ M, ($FeSO_4$) $= 3 \times 10^{-5}$ M, (H_2O_2) $= 4.89 \times 10^{-2}$ M and pH=3. The result of photocatalytic bleaching of Reactive Orange-16 is graphically presented in Fig. 2.

It was observed that absorbance decreases with the increase in time of irradiation indicating that the dye is degraded on irradiation. A graph between 2 + log Abs. and

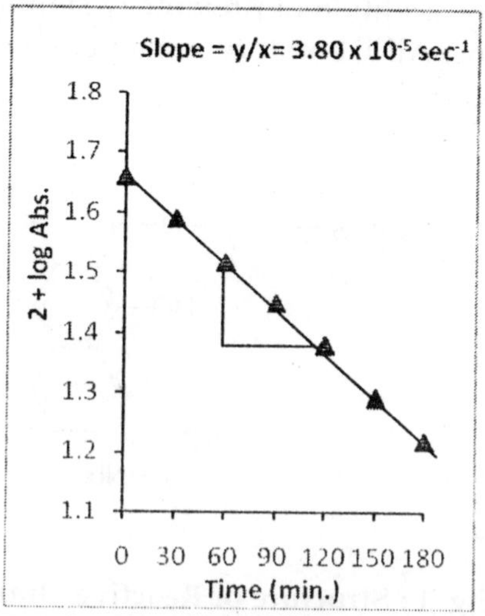

Fig. 2: A plot showing a typical run of photobleaching of Reactive Orange-16 by Photo-Fenton reagent under the optimized conditions.

time has been plotted. The linearity of the plot indicates that the photocatalytic bleaching of Reactive Orange-16 follows a first order kinetics. The rate constant of this photobleaching process was determined using the expression.

Rate (k) = 2.303 x Slope

= 8.75×10^{-5} sec^{-1}

The effect of variation in various reaction parameters has been studied, e.g. pH, concentration of the dye, concentration of $FeSO_4$ and concentration of H_2O_2.

Effect of variation in hydrogen ion concentration on the rate of decolorization of the dye

Fenton oxidation is known as a highly pH dependent process since pH plays an important role in the mechanism of $^{\bullet}OH$ production in the Fenton's reaction. At high pH (pH > 3), the generation of $^{\bullet}OH$ gets slower because of the formation of the ferric hydroxo complexes.

On the other hand, at very low pH (< 2.0), hydrogen ions acts as $^{\bullet}OH$ radical scavengers. The reaction is slowed down due to the formation of complex species $(Fe(H_2O)_6)^{+2}$, which reacts more slowly with peroxide compared to that of $(Fe(OH)(H_2O)_5)^{+2}$.

In this study, photodegradation was performed at different pH from 1 to 4.5. Fig. 3 also prove that pH 3 is the optimal pH for RO16 decolorization. Hence all subsequent experiments were carried out at pH 3.

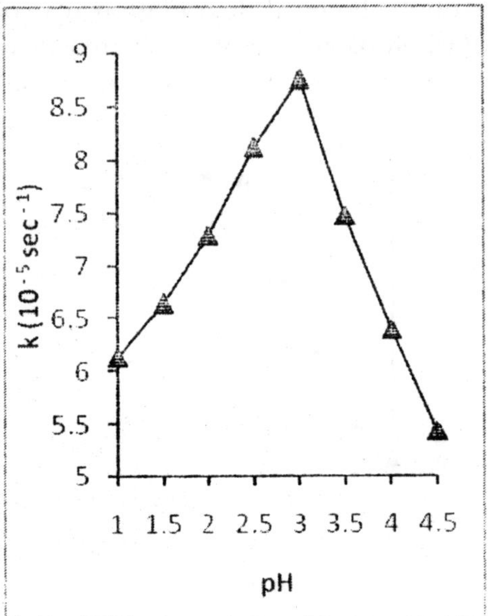

Fig. 3: A plot showing effect of variation in hydrogen ion concentration on the rate of decolorization of the dye by Photo-Fenton reagent.

Effect of variation in dye concentration on the rate of decolorization of the dye

The effect of dye concentration on the degradation of Reactive Orange-16 was studied at different concentrations varying from 1.0×10^{-5} M to 8.0×10^{-5} M keeping all other factors identical. The result (Fig. 4) reveals that the rate of photobleaching of dye decreases with the increase in the concentration of the dye. The reason behind it is that the increase in the initial concentration of the dye lies in consistency of the hydroxyl radicals concentrations for all the dye molecules and therefore the rate of decolorization decreases.

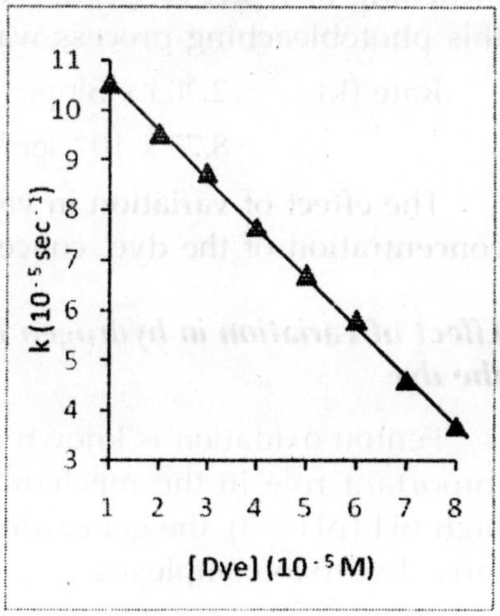

Fig. 4: A plot showing effect of variation in dye concentration on the rate of decolorization of the dye by Photo-Fenton reagent.

Effect of variation in catalyst concentration on the rate of decolorization of the dye

Keeping all other factors identical, the concentration of catalyst was changed and its effect on the rate of photochemical degradation was observed. The result of Fig. 5 reveals that the rate of photobleaching of dye increases with the increase in the concentration of catalyst $FeSO_4$ up to 3×10^{-5} M. The increase in ferrous ions in the

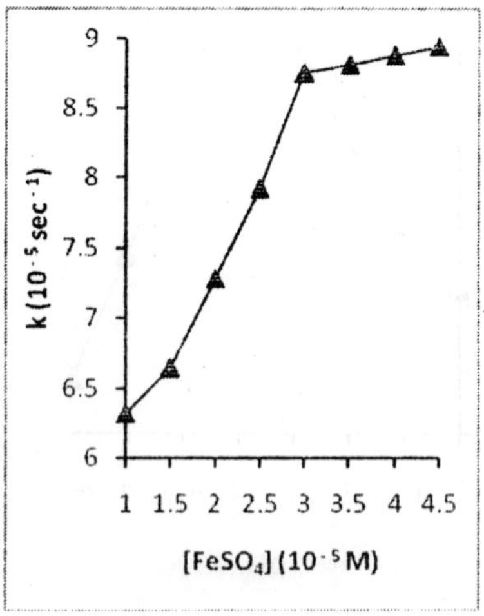

Fig. 5: A plot showing effect of variation in catalyst concentration on the rate of decolorization of the dye by Photo-Fenton reagent.

reaction mixture is accompanied by enhanced generation of $^{\bullet}OH$ radicals, consequently increasing the rate of photodegradation. After the optimal Fe^{2+} addition, the higher dose of Fe^{2+} resulted in a brown turbidity that causes the recombination of $^{\bullet}OH$ radicals and Fe^{2+} reacts with $^{\bullet}OH$ as a scavenger. Therefore, on further increase, the rate becomes almost constant.

Effect of variation in hydrogen peroxide concentration on the rate of decolorization of the dye

The effect of H_2O_2 concentration on the degradation of RO16 is shown in Fig. 6. The result reveals that the rate of photobleaching of dye increases with the increase in the concentration of H_2O_2 upto 4.89 x10^{-2} M. This can be explained on the basis that at higher concentration of H_2O_2, more hydroxyl radicals are produced which degrade more dye molecules.

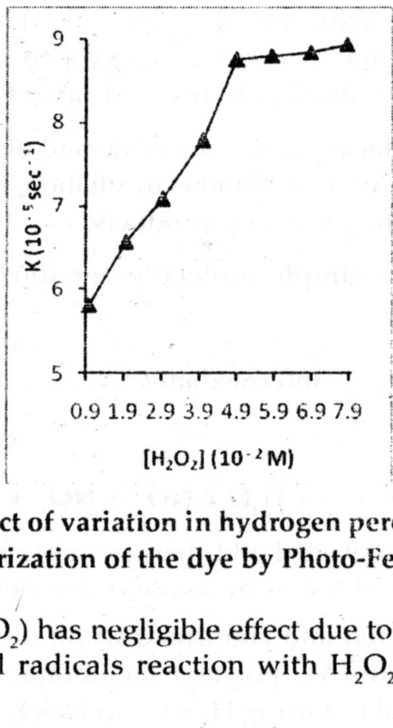

Fig. 6: A plot showing effect of variation in hydrogen peroxide concentration on the rate of decolorization of the dye by Photo-Fenton reagent.

Further increase in (H_2O_2) has negligible effect due to the recombination of $^{\bullet}OH$ radicals and also hydroxyl radicals reaction with H_2O_2 contributing to the $^{\bullet}OH$ scavenging capacity.

$$^{\bullet}OH + {}^{\bullet}OH \longrightarrow H_2O_2$$

$$^{\bullet}OH + H_2O_2 \longrightarrow {}^{\bullet}O_2H + H_2O$$

$$^{\bullet}O_2H + {}^{\bullet}OH \longrightarrow H_2O + O_2$$

MECHANISM

The Fenton reagent produces $^{\bullet}OH$ radicals by interaction of H_2O_2 with Fe^{2+} ions (Eq: 1).

$$Fe^{2+} + H_2O_2 \longrightarrow Fe^{3+} + {}^{\bullet}OH + OH^{-} \quad (1)$$

In dark, the reaction is retarded after complete conversion of Fe^{2+} to Fe^{3+}. But in presence of light, photoreduction of Fe^{3+} to Fe^{2+} takes place which produces new $^{\bullet}OH$ with H_2O (Eq: 2).

$$Fe^{3+} + H_2O + h\nu \longrightarrow Fe^{2+} + {}^{\bullet}OH + H^{+} \quad (2)$$

Ferrous ions will undergo oxidation to ferric ions by the addition of $^{\bullet}OH$ radicals, while ferric ions are reduced to ferrous ions by the incorporation of $^{\bullet}OOH$ radicals producing H^{+} ions (Eq: 3, 4).

$$Fe^{2+} + {}^{\bullet}OH \longrightarrow Fe^{3+} + OH^{-} \quad (3)$$

$$Fe^{3+} + {}^{\bullet}O_2H \longrightarrow Fe^{2+} + O_2 + H^{+} \quad (4)$$

$^{\bullet}OOH$ radicals are highly unstable in water and undergo facile disproportionation rather than reacting slowly with the dye molecules. The participation of the $^{\bullet}OH$ radical as an active oxidizing species was confirmed using the hydroxyl radical scavenger 2-Propanol, which drastically reduced rate of degradation.

The hydroxyl radical attacks on dye molecule and abstracts a hydrogen atom or adds itself to double bonds.After continuous irradiation, the complete mineralization of dye occurred via converting into end products.

The end products are simple molecules or ions and less harmful to the environment (Eq: 5).

$$\underset{\text{Reactive Orange-16 (S)}}{C_{20}H_{17}N_3Na_2S_3O_{11}} \xrightarrow{OH^{-}} \text{Intermediates} \longrightarrow \underset{(P)}{\text{End Products}} \quad (5)$$

End products $\equiv CO_2 + H^{+} + H_2O + NO_2^{-} + NO_3^{-} + SO_4^{2-}$

The end products were detected and their presence in the reaction mixture was ascertained either by chemical test or by ion selective electrode method.

Nitrate ions were detected and confirmed using nitrate ion selective electrode which is having a solid-state PVC polymer membrane. Nitrite ions were detected and confirmed by a chemical test using H_2SO_4 and $FeSO_4$. A positive test for presence of nitrite ions is indicated by a dark brown solution, arising from the iron nitric oxide complex. Sulphate ions were detected and confirmed by gravimetric analysis in which excess of barium chloride solution was used and sulphate ions are precipitated as $BaSO_4$. CO_2 was confirmed by introducing the gas to freshly prepared limewater. The test solution (limewater) turns milky which indicates its presence.

The whole process is picturised (Fig.7).

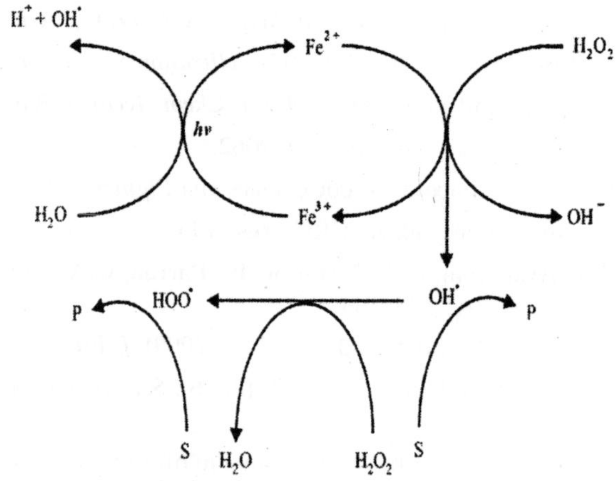

Where S = Dye and P = End Products

Fig. 7: A Schematic representation of Photo-Fenton Chemistry

CONCLUSIONS AND FUTURE PERSPECTIVES

Photo- Fenton process was used for the photo-catalytic degradation of dye reactive Orange 16. Based on the results, it can be concluded that the decoloration of dye is strongly dependent on various reaction parameters e.g. pH, concentration of the dye, concentration of $FeSO_4$ and concentration of H_2O_2.

The photocatalytic degradation of Reactive Orange-16 was observed at 490 nm. The optimum conditions for the photobleaching of dye were (Dye) $=3\times10^{-5}$ M, ($FeSO_4$) $=3 \times 10^{-5}$ M, (H_2O_2) $= 4.89 \times 10^{-2}$ M and pH=3. Kinetic studies reveal that photocatalytic bleaching of Reactive Orange-16 follows a first order kinetics.

The great interest of the academic community for the use of AOPs in waste water treatment is reflected by a number of publications that have been produced during the last decade. Major drawbacks of some methods is the use of costly chemicals, high operational cost, increased energy consumption as well as the formation of unknown intermediates which in some cases could be more toxic than the parent compounds. Based on the above, the main future challenges for using AOPs in waste water treatment could be the development of efficient and low cost materials to promote sufficient treatment, the use of renewable energy sources, the targeting of new classes of pollutants and the commercialization of processes which have been so far used in the laboratory.

REFERENCES

Bag, B.C., Sai, M., Sekhar, K. and Bhattacharya, C. (2009). *J. Environ. Sci.*, 3: 34.

Chatterjee, D. and Dasgupta, S. (2005). *J. Photochem. Photobiol.*, 6: 186.

Gabriel, H. and Hong (2008). *J., Research J. App. Sci.*, 3: 216.

Haperman, C. J., Anderson, B. G., Torrents, A. and Acher, A. (1997). *J. Aric. Food Chem.*, 45: 1006.

He, Z., Song, S., Zhou, H., Ying. H. and Chen J. (2007). *Ultrason. Sonochem.*, 14: 298.

Juang, R.S., Tseng, R.L., Wu, F.C. and Lee, S.H. (1997). *J. Chem. Technol. Biotechnol.*, 70: 391.

Kavitha, V. and Palanivelu, K. (2005). *Water Res.*, 39: 3062.

Kusic, H., Bozic, A. L. and Koprivanace, N. (2006). *Dyes and Pigments*, 1.

Ming-Ho, Yu. (2005). *Environmental toxicology*, CRC Press LLC.

Orozcoa, S.L., Bandala, E.R, Aranclbia, C.A, Serranoc, B., Parraa, C.A. and Perezd, I.H. (2008). *J. Photochem. Photobiol. A: Chemistry*, 198: 144.

Segura, C., Zaror, C., Mansilla, H.D. and Mondaca, M.A. (2009). *J. Environ. Sci.*, 3: 34.

Solmaz, S. K. A., Birgul, A., Vstun, G. E. and Yonor, T. (2006). *Society of dyers and colourists, color. Techno.*, 122: 102.

Swaminathan, K., Sandhya, S., Caramlin, A., Sophia Pachhade, K. and Subrahmanyam, Y.V. (2003). *Chemosphere*, 50: 619.

Ugurlu, M. and Kula, I. (2007). *Environ. Sci. Poll. Res.*, 14: 319.

Yang, C. and Mc Garrham, J.J. (2005). *J. Hazar. Mater., B*, 127: 40.

Zollinger, H. (2003). *Color Chemistry, Wiley- VCH*, Weinheim, Germany.

10

Urbanization in Non-human Primates and Conflict with Human Being

Goutam Sharma, Chena Ram and L.S. Rajpurohit

Introduction

Non-human primates, comprising monkeys, langurs, apes, lemurs and lories, etc. inhabit most part of the biodiversity rich, 4 major biogeographically tropical and subtropical regions of the world. They perform ecological services such as seed dispersers, pollinators, primary consumers and as food for top predators. They are therefore, good indicators of the general health of the ecosystem and are helpful in conservation planning. Non-human primates are the only appropriate animal models in many basic and applied studies because other species are not susceptible to the disease or disorder under study. They are invaluable for biomedical research because of their marked similarities to humans in almost all aspects of their anatomy, physiology, neurology, endocrinology, immunology and behaviours.

The nonhuman primates are represented with 63 genera and about 600 species or subspecies in about 92 countries, Brazil topping in the list with 77 species. A total of 21 species are recorded from the Indian subcontinent.

India has an exceptionally rich heritage of non-human primate population altogether 16 species and 39 subspecies. This richness exhibits the biological and environmental diversity of India, ranging from mountain habitats in the Himalayas, the deserts of Rajasthan, agricultural plains of the Gangatic basin, subtropical forests of the northeast, mangrove estuaries of coastal India, tropical forests and the coral reefs of the south India.

Since man evolved from the non-human primates (monkeys, apes and allies), thought some two million years ago, and is indeed one of them. Although, human being is the highest, it is both natural and logical from the evolutionary point of view to seek on understanding of human behavior, especially in social relationship. One of the most significant developments in human relationships in the last twelve thousand years, after the advent of agriculture and especially after the growth of industrialization has been the aggregation of vast numbers of humans in cities. And this process called urbanization has brought in its own problems. Simultaneously with the development of human urban settlements, a few non human primate species have also become urbanized, with the difference that while man created his own city environment by building homes and other requisite, the non-human primate species which took to urbanization simply usurped the existing either ruined or living human dwelling facilities and adopted themselves to live in such new environment. This paper concerns the new characteristics acquired by urbanized primate populations and compares them with the unurbanized counter parts (i.e. rural and forest-dwellers) and with man in aspects such crime/infanticide, cannibalism etc.

Urbanization in Non-human Primates

Man-monkey association is as old as man's own existence. Of nearly 225 living species of non-human primates, three Indian species have become urbanized. They are the rhesus macaque (*Macaca mulatta*), the bonnet macaque (*M. radiata*) and the common/Hanuman langur (*Semnopithecus entellus*). The rhesus macaques live in northern India. The bonnet in south and the common langur is widely distributed and the most adaptable species found all over the Indian sub-continent. Both the macaques, however are so thoroughly urbanized that they are as entirely at home in cities as they are in their original forest homes. The Hanuman langur have also developed fairly close relationships with man, but have not become truly urbanized.

In the Indian context the man-monkey relationship is remarkable. On one side people consume blood and flesh of monkeys as medicines, trap, kill and eat them as food, on the other side people keep them as pets, trained them to play , feed and protect them. Urbanized populations are provisioned frequently due to religious sentiment of people. So human attitude towards monkey differ from area to area and species to species. For example, many tribals living in the interior of the forest trap, kill and eat monkeys irrespective of the nature of species. Likewise, monkeys are not liked in the areas of massive agriculture, horticulture and other plantations since they said and damage the crops and orchards. In such areas they are considered pests. (Roonwal and Mohnot, 1977). In yet another situation monkeys have become commensalisms and are competitors of human being in and around villages, towns and cities. These are "Urbanized monkeys".

The Rhesus Macaque *(Macaca mulatta,* Zimmermann)

The rhesus macaque, commonly known as 'Bandar' is medium-sized monkey with a stout body and a relatively short tail. It occurs in South and South East Asia- from eastern Afganistan via Pakistan, Northern India, Nepal, Bangladesh and Mymnar, East to Thialand and Vietnam and small pockets in Southern Tibet. Its sociobiology has been studies by a number of researchers (Kaufmann, 1967, Lindburg. 1971). It lives in groups composed of 8-180 members. These groups are generally bisexual (i.e. comprising adults of both sexes) and multimale (having several adult males). There is a central and a peripheral grouping of males, the former being dominant over the later. A dominate hierarchy among males occurs, with an alpha or the most dominant male and two or more less dominant ones. Females also show sank order among themselves.

In many of the northern Indian cities, any observer can easily convinced that how the rhesus macaque has become completely urbanized. It has entirely at home on roof tops, in temples and in railway yards. In the present study we are more concern to know that if any difference between city-dwellers and forest-dwelling monkeys, what they are and how. The natural behavior changes take place. It has been observed that the urban and forest-dwelling groups differ markedly in food preferences, foraging and roosting habits, behaviors at the feeding site, aggression, curiosity etc.

The forest-dwellers live principally on leaves and fruits while urban monkeys on cooked human food (chapaties, sweets, roasted gram, groundnut etc.) as well as on fruit and vegetable stolen from shops and houses. Forest-dwellers obtain food naturally and fearlessly in nature where it is easily available. The city-dwelling animals have to steal or pilfer it from food shops, vegetable and fruit shops in open market and domestic kitchens. They are constantly in tension with citizens who chase them away and many of times beat them up. However, at the same time, the religious-minded people provision them fruits and cooked food. And therefore, the city monkeys are alert, suspicions and bold.

The roosting sites are not fixed for forest-dwellers, they have some preferred sites but they can roost for the night in any convenient tree near which happen to be at the end of the day. On the other hand, city dwellers have fixed sleeping sites, such as ruined houses and temple roof-tops. At the feeding place, city –dwelling monkeys show less aggression among them selves, but are basically more aggressive in comparison to forest-dwellers. The greater basic aggressiveness of the urban monkeys may have evolved due to the more restrictive nature of the urban environment as regards space and the limited food supply, resulting in greater competition among themselves and man. The forest-dwellers are observed to be very shy, retiring nature and are afraid of man, when trapped and put in cages, while city-dwellers are bold and are not afraid of human proximity. The former show less curiosity and less desire to explore than the city-dwellers.

The Bonned Macaque (*Macaca radiate, E. Geoffroy*)

The bonnet macaque is a south Indian monkey which is found in the whole of peninsular India north to Satara. Its crown hair are arranged like a cap or bonnet on the head, with the hairs radiating in all directions. It is smaller in body size than rhesus macaque and has a relatively longer tail, almost as long as the head and body. The sociobiology of this species has been studied by several workers (Parthasarathy,

1972, Sugiyama, 1971, Rahaman and Parthasarathy, 1969, Simonds, 1975). The bonnet macaque prefers rural and sub-urban areas with tall banian and other trees and is not an animal of the deep, thick forest. However it is equally at home in cities, temples and railway yards, and is the southern Indian counterpart of the north Indian rhesus monkey. The food habit of bonnet macaque in both forests and cities are similar to those of the rhesus. It lives in groups of some 7-80 individuals, groups are multimale and bisexual. Occasionally, small all-male groups and solitary males are found. The differences in habits and behaviors of city-dwelling and forest groups of bonnet monkeys also more or less similar to the rhesus macaque.

The Hanuman Langur (*Senopithecus entellus,* Dufresne)

The Hanuman langur or common langur is large, sleek, black-faced, long tailed monkey found throughout the length and breadth of India, and spreads over into southern Tibet and Nepal in the north, Bangladesh and Mymnar in the east and Srilanka in the south. Due to having long tail, it's name 'langur' or 'langoor'.

The animal has been well-studied by reiveral researchers for over last 40 years (Jay, 1963, Sugiyama, 1965, Mohnot, 1974, Hrdy, 1974, Winkler, 1981, Sommer, 1985, Agoramoorthy, 1987, Rajpurohit, 1987, Borries, 1989, Srivastava, 1989, Bhaker, 2002, Sharma, 2007). The common langur lives in small to large groups varying in size from 7 to abort 125 individuals, Groups are generally of the bisexual type and are either unimale or multimale, All male bands are also found. It lives in all kinds of ecological situations, ranging from dense and hill forests to open scrub, and open, rocky, xerophytic areas to rain forests, from sea level to high attitude of about 4000 m in the Himalayas. Although many groups live on the outskirts and within city areas of Jaipur, Jodhpur and Ajmer, They have not become truly urbanized as the rhesus macaque. Langurs also raid regularly the houses, roof tops, crop fields, orchads or kitchen gardens in the urban environment. At many of the sites, the religious-minded citizens regularly provision the langur with cooked food and fruits and vegetables. They have developed close relationship with man and are fearless.

Infanticide in Langurs

Although isolated cases of infanticide have been reported in some 6-7 species, including the prosimians, macaques, baboons and chimpanzees (here act sometimes involving cannibalism) only in the Hanuman langur, that infanticide has been observed as a regular and normal feature in natural populations (Hrdy, 1974, Sugiyama, 1965, Sharma et al, 2010) A newly changed female, snatches away the infant, bites it viciously and leaves it dying. The mother, and occasionally one or two other females of the group, try to chase away the attacking male, sometimes succeed. The attack is premeditated and is not mounted on the spur of the moment. For days and sometimes weeks, the male stalks the intended victim and tries out the attacks, until he succeeds. The infants so killed are usually under 3 months old, but sometimes on yearling is the victim. Infanticide seems to sexually stimulate the mother who comes into heat a few days after the event and copulates with the killer male or with other males. However not all the new resident indulge in this practice, about 40% of the new residents have been observed non-infanticide (Rajpurohit, 1987). The sexual hypothesis also not completely valid as male has not been observed killing only male infants.

The Urbanization Process

Any adaptation in primates inchiding man that may have arisen in response to true urbanization must have been acquired within the last 10-12 thousand years or so, when man started becoming urban as cities began to arise. The urbanization of the only known two fully urbanized non-human primates (i.e. rhesus and bonnet mecaques) took place in India. These two, and a third the Hanuman langur (also Indian) have become an intimate part of the Hindu culture of tolerance (Roonwal, 1977). It is obvions that, within a species urbanization did not occur just once, but several times (and step by step). As a small village become a town and a city, neighbouring groups gradually become adopted to live in urban conditions was acquired new habits and behavior patterns. This process was repeated for another city and for new population graups, and so on. Thus, for example primate urbanization in Jaipur cannot be more than 280 years old (the city was founded in 1728) and in Jodhpur it cannot be more than 500 years old (as Jodhpur city was founded in 1459 ad by Rao Jodha), whereas in ancient cities like Ayodhya and Varanasi, it could well have begun a thousand years ago or earlier.

The so called urbanized non-human primates (i.e. rhesus and bonnet meaques and Hanuman langurs) are always a nuisance, and also damage crops and orchards. The Hindu belief in the sacredness of all life and the weaving of monkeys into ancient Hindu mythology and literature have helped to create a climate of tolerance. Man has of course, played a direct and important role. If he wishes, can easily exterminate macaques and langurs in cities. But he would rather allow them to wander the streets as if they own these, or suffer frequent loss of vegetables and fruits from his shops, or put gauze doors his kitchen to prevent monkeys getting in. However, all primates do not have the same capacity to become urbanized. This is clear in case of Hanuman langur. For centuries it has had the some opportunities as the rhesus and the bonnet macaques, but has never become fully urbanized as these two macaque species.

Conflict with Man

Man has created for himself an ecological setting, as well as social and psychological settings, to suit his physical and mental needs. Some of these creations have gone out of control (e.g. overcrowding, insanitation), but being highly adaptable, man continues to tolerate them and even thrive in them. The result is overpopulation. Such a state has not yet arisen in the urbanized non-human primates. They have also acquired new adaptations, and for instance, no increase in fighting, and no sign of crime or other abnormal types of behaviors. There is also no marked change in social organization or sexual habits. Both wild and urbanized macaque females come into estrus only at certain time of the year and will receive male only then, a fact which restrains excessive breeding. In contrast, women are receptive and fertile the year round, an ability which results in greatly increased breeding capacity. This fact also implies for Hanuman langur living in and around Jodhpur. The Human population increased from 150000 in 1977 to about 10,00,000 in 2005, while the langur population here become 1900 animals in 2005 from 850 in 1977 (Rajpurohit et. al. 2006). Whether the urbanized macaques and langurs may, with food abundance and social security, acquire an year-round sexual capacity in the next ten thousand years is a matter for speculation.

Human population growth and activities like deforestation, agriculture and urbanization lead to an ever-increasing encroachment on wildlife habitats. Reduction of wild animals' natural habitats forces species unable to adapt to altered habitats into small marginal patches. Observations have been scored who initiated the interaction (human or Hanuman langurs), the noted age classes and sex of the human and the langur. Conflict with non-human primates is related to levels of tolerance, a dynamic that is influenced by the degree of disturbance to pursuance of human activities caused by the animals. There can thus be no predetermined number of non-human primates that can be predicted to cause conflict. Non-human primates display aggression towards humans primarily over food (crops, tourist feeding, etc.); non-human primates become bold when they have learned that an easy meal can be won by showing aggressiveness towards humans. From a non-human primates' point of view a human engages in an act of subservience towards it by giving it food - and it may follow up with aggression if food is withheld or not given. In short, by feeding non-human primates humans are teaching them to be aggressive towards humans. They may also become aggressive when they perceive that a human is threatening one of their youngsters.

Human Impact on Nonhuman Primates

There are large number of areas where in nonhuman primates and humans impact one another in significant ways. This zone of interconnection is largely theoretically and practically excluded from most examinations of socioecology/ behavioral ecology of human and nonhuman primates. For example, traditional socio-ecological models used in assessing primate behavior rely on: female distribution, patterns of competition (scramble vs. Contest), presence/distribution of predators, resource availability, and forest structure/composition. These elements can be substantially impacted by a variety of anthropogenic processes.

This in fact may or may not be a factor, for example in Hanuman langurs

(*Semnopithecus entellus*) for which there are a variety of populations where infanticide is reported and others where it is not. This is not to say that infanticide necessarily arises from the human population induced stress, but rather that differential interactions between humans and the habitat (and monkeys) may in fact create diverse social, demographic and ecological conditions that make infanticide more or less viable as a strategy and/or as an occurrence. Therefore, one could compare elements of the human communities and their habitat use patterns across different langur sites to see of there are any correlates with langur behavior (such as infanticide).

In an opposite context from hunting, food enhancement (via direct provisioning, incidental provisioning, crop raiding, house garden raining, garbage exploitation, plantation living, etc...) can facilitate a reduction of food stress and at the same time increase risk of attack (by humans/dogs/etc) creating different patterns of stress response (chronic vs. acute). Additionally, the foods being exploited can alter the metabolic patterns of nonhuman primates. These factors can then influence dominance hierarchies and behavioral profiles of a nonhuman primate group.

Nonhuman Primates Impact on Humans

Human conceptualizations of nature influence, and are influenced, by ecological patterns and the nonhuman co-inhabitants of shared ecologies/habitats. In this context, nonhuman primates may play substantial roles. In areas of sympatry many human communities keep primates as pets.

In addition to the impacts on cultural patterns and perceptions, the spatial overlap and interactions between humans and nonhuman primates can create an environment where more than space is shared; there may also be co-mingling of infectious agents. While it is generally assumed that human-non-human primate infectious agent "sharing" has deleterious repercussions, our understandings of the patterns and contexts of these shared environments and their evolutionary implications remains rather incomplete (Jones-Engel et al., 2005).

Close contact and range overlap between humans and nonhuman primates does introduce a very real and potentially dangerous situation of disease transmission (Engel et al., 2002, Jones-Engel, et al., 2005). In these cases, humans risk exposure to a number of simian viruses, including simian T cell lymphotropic viruses (STLV), simian retrovirus (SRV), simian foamy virus (SFV) and Herpes B virus in addition to other known infectious agents (Engel et al, 2002, Jones-Engel et al., 2005, Wolfe et al., 2004). It is also important to point out that nonhuman primates are potentially at risk from human pathogens as well, including measles, influenzas and other respiratory pathogens such as parainfluenza and tuberculosis (Jones-Engel et al., 2001).

Survival and reproduction of individuals depend on their ability to locate and harvest sufficient food to meet their nutritional needs. All primate species interact with a variety of food distributed in their home range, which is within their reach.

In the proposed research crop raid by Hanuman langur is proposed. Their ability to learn things very quickly and change in the behaviour accordingly, makes them very successful and potential crop raiders. Crop damage is very common along the immediate periphery of wildlife sanctuaries and national parks, by primates and other wild animals. Cultivated food plants and their seasonality in the diet of Hanuman langur. Langurs are highly adaptive animals and feeding upon variety of food items, which includes, natural, cultivated and artificial food. These animals are capable of damaging most agricultural and horticultural crops.

In Sri Lanka langurs eats young leaves, shoots, buds of mangoes and other fruits and they some time killed as crop raider (Muckenhirn, 1972). But, in India peoples are not killing langurs mainly because of prevalent religious sentiments and because of Wildlife Protection Act, 1972. The langur is regarded as sacred by Hindus, who often feed them, generally every Tuesday, which is the traditional day of the monkey-god- Hanuman (Roonwal And Mohnot, 1977). But farmers frequently threaten them, throwing stones, whistle, shout at them and use firecrackers to scare them away from the field. But, langurs are very intelligent and well organized when they raid crop fields.

Human-wildlife conflicts also undermine human welfare, health and safety, and have economic and social costs. Nuisance encounters with small animals, exposure to zoonotic diseases, physical injury or even death caused by large predators' attacks have high financial costs for individuals and society in the form of medical treatments to cure and prevent infections transmitted from animals through human contact. Such broad environmental, human health and safety, economic and social impacts suggest that governments, wildlife managers, scientists and local communities need to recognize the problem and adopt measures to resolve it in the interest of human and environmental well being.

Summary

Three of some 225 species of non-human primates, the rhesus macaque (*Macaca mulatta*), the bonnet macaque (*Macaca radiata*) and the common Hanuman langur (*Semnopithecus entellus*) have become urbanized. All three species are Indian- the rhesus macaque lives in northern India, the bonnet macaque in south and the common langur is widely distributed all over Indian subcontinent. In these non-human primates species, behavior differences have developed between urbanized population and forest-dwellers. The observation suggest that city-dwellers are more explorative, bolder, more cunning, more particular in setting with permanent rousting, more aggressive and more tolerant in food-sharing than the forest-dwellers.

Non-human primates are on the whole peaceful, and peace is largely achieved between groups by each group having its more or less well defined home range or territory, and within a group by a well established system of dominance hierarchies. Crime is generally not seen in non-human primates. In an exceptional case, in the

Hanuman langur (*Semnopithecus entellus*) and Chimpanzee (*Pan troglotydis*), infanticide by males is a normal feature. However, the city-dwelling langurs are comparatively less aggressive than the urbanized rhesus macaques. The process and consequences of urbanization are analyzed and discussed to compare with man. The growth of cities has occurred only within the last 15-20 thousand years, after the advent of agriculture and urbanization is probably not more than 10 to 12 thousand years old. And non-human primate urbanization has also occurred in that period. Study of Hanuman langur in Jodhpur suggests that urbanization may take place fully in a short period of abort 100 years.

Acknowledgement

The authors are grateful to Prof. S. M. Mohnot chairman, Primate Research Center, Jodhpur for continuous encouragement. To prof. Devendra Mohan Head , Department of Zoology, J.N.V. University, Jodhpur for providing facilities and logistic support during this study.

References

Agoramoorthy, G. (1987). Reproductive behaviour in Hanuman langur, *Presbytis entellus*. Ph.D. Thesis. University of Jodhpur, Jodhpur, India.

Bhaker, N.R. (2002). Role of vocal communication in sociobiology of Hanuman langur, *Semnopithecus entellus entellus* around Jodhpur (India). Ph.D. thesis. J.N.V. University, Jodhpur.

Borries, C. (1989). Konkurrenz unter freilebenden Langurenweibchen (*Presbytis entellus*). Dissertation, Gottingen, Gerorg-August Universitat.

Engel, G.A., L. Jones-Engel, M.A. Schillaci, K.G. Suary-ana, A. Putra, A. Fuentes and R. Henkel. (2002). Human exposure to herpesvirus B-seropositive macaques, Bali, Indonesia. Emerg. Infect. Dis. 8: 789–795.

Hrdy, S.B. (1974). Male-male competition and infanticide among the langurs (*Presbytis entellus*) of Abu, Rajasthan, Folia Primatol.; 22:19–58.

Jay, P.C. (1963). The social behaviour of the langur monkey. Ph.D. thesis, University of Chicago, Illinois.

Jones-Engel, L., G.A. Engel, M.A. Schillaci, A. Rompis, A. Putra, K.G. Suaryana, A. Fuentes, B. Beer, S. Hicks, R. White, B. Wilson and J.S. Allan (2005). Primate-to-human retroviral transmission in Asia. Emerg. Infect. Dis. 11: 1028.

Jones-Engel, L., G.A. Engel, M.A. Schillaci, R. Babo and J. Froehlich (2001). Detection of antibodies to selected human pathogens among wild and pet macaques (*Macaca tonkeana*) in Sulawesi, Indonesia. Am. J. Primatol. 54: 171–178.

Kaufmann, J.H. (1967). Social communication among primates (Ed. S.A. Altmann), Chicago (University of Chicago Press), pp. 72-88.

Lindburg, D.G. (1971). Primates Behaviour: Development in Field and Laboratory Research. Vol. 2 (Ed. By L.A. Rosenblum). New York, Academic Press.

Mohnot, S.M. (1974). Ecology and Behaviour of the Common Indian Langur, *Presbytis entellus*. Ph.D. thesis, Univ. of Jodhpur, Jodhpur.

Muckenhirn, N.A. (1972). Leaf-eaters and their predators in Ceylon: Ecological role of gray langurs, *Presbytis entellus* and Leopards. Ph.D. thesis, University of Maryland.

Parthasarathy, M.D. (1972). Proceeding of the 4th International congress on primatology (IPS, Cambridge), p.56.

Rahman, H. and M.D. Parthasarathy (1969). Studies on the social behaviour of bonnet monkeys. Primates; 10: 149-162.

Rajpurohit, L.S. (1987). Male social organization in Hanuman langurs (*Presbytis entellus*). Ph.D. thesis, Univ. of Jodhpur, Jodhpur.

Rajpurohit, L.S; Chhangani, A.K. and. Mohnot, S.M. (2006). Population dynamics of Hanuman langur, *Semnopithecus entellus* around Jodhpur (India) during 1995-2001). PROC. NAT. ACAD. SCI. INDIA, Vol. 76 (B), II: 141-147.

Roonwal, M.L. & Mohnot, S.M. (1977). Primates of South Asia, Harvard University Press, Cambridge, Mass.

Sharma, G. (2007). Paternal Care in Hanuman langur, *Semnopithecus entellus entellus* around Jodhpur (India). Ph.D. thesis. J.N.V. University, Jodhpur.

Sharma, G., Chena Ram and Rajpurohit, L.S. (2010). A Case Study of Infanticide after Resident Male Replacement In *Semnopithecus entellus* around Jodhpur (India). PROC. OF ZOOLOGICAL SOCIETY, Vol. 63 (2),: 93-98.

Simonds, P.E. (1975). Cited in Primate behaviour: field steadies of monkeys and apes, edited by I. De Vore (Holt, Rinehart and Winston Inc., New York).

Sommer, V. (1985). Weibliche and Mannliched reproduction strategien der Hanuman Languren (*Presbytis entellus*) von Jodhpur, Rajasthan Indien: Dissertation. Gottingen: Georg August University.

Srivastava, A. (1989). Feeding Ecology and Behaviour of Hanuman langurs, *Presbytis entellus*. Ph.D. thesis. University of Jodhpur.

Sugiyama, Y. (1965). On the Social Change of Hanuman Langurs (Presbytis entellus) in their natural conditions. Primates; 6: 213-47.

Winkler, P. (1981). Zur oko-ethologe freilehender Hanuman languren (*Presbytis entellus entellus*) Dufresen, 1797) in Jodhpur (Rajasthan), Indien. Ph.D. thesis, Goettingen, George-August Universitat.

Wolfe, N.D., W.M. Switzer, J.K. Carr, V.B. Bhullar, V. Shanmugam, U. Tamoufe, A.T. Prosser, J.N. Torimiro, A. Wright, E. Mpoudi-Ngole, F.E. McCutchan, D.L. Birx, T.M. Folks, D.S. Burke and W. Heneine (2004). Naturally acquired simian retrovirus infections in central African hunters. Lancet 363: 932–937.